MAKING ARMS
IN THE
MACHINE AGE

MAKING ARMS
IN THE
MACHINE AGE

Philadelphia's Frankford Arsenal,
1816–1870

James J. Farley

The Pennsylvania State University Press
University Park, Pennsylvania

Library of Congress Cataloging-in-Publication Data

Farley, James J., 1942–
    Making arms in the Machine Age: Philadelphia's Frankford Arsenal, 1816–1870 /
James J. Farley.
      p.   cm.
    Includes bibliographical references and index.
    ISBN 0-271-01000-2
    1. Frankford Arsenal (Pa.)—History.   2. Frankford (Philadelphia, Pa.)—History.
I. Title.
    UF543.P6F37   1994
    338.4'7623442'0974811—dc20   93-19126
                                                           CIP

Published by The Pennsylvania State University Press,
Barbara Building, Suite C, University Park, PA 16802-1003

It is the policy of The Pennsylvania State University Press to use acid-free paper for the
first printing of all clothbound books. Publications on uncoated stock satisfy the minimum
requirements of American National Standard for Information Sciences—Permanence of
Paper for Printed Library Materials, ANSI Z39.48–1984.

Contents

List of Illustrations

Acknowledgments

If this acknowledgment listed all the people who contributed to this study, it would be longer than the work itself. The wealth of archival material in the Philadelphia area is a treasure trove for historians. Robert Plowman of the Mid-Atlantic Branch of the National Archives unearthed and helped me to make sense of the records of the Frankford Arsenal and several early nineteenth-century censuses on which, to a large extent, this study is based. Linda Stanley of the Historical Society of Pennsylvania, Jefferson Moak of the Office of the Archives of the City of Philadelphia, Ken Finkle of the Library Company of Philadelphia, and Sally Elk of the City of Philadelphia Historical Commission assisted in similar fashion.

Mike Zuckerman, John Murrin, Judith McGaw, Richard Dunn, Robert Gordon, Carolyn Cooper, Wayne Bodle, and all of the other participants in the Philadelphia Transformation Project and its parent, the Philadelphia Center for Early American Studies, helped to fashion this work. The center is a jewel of great value for historians of early America; it allows us to meet, share, and sometimes challenge — but always understand — each other's work.

A large urban university such as Temple University has many outstanding departments, but its history department ranks among the finest. Herbert Ershkowitz patiently guided this study from inception to conclusion; Allan Davis and Russell Weigley contributed both their knowledge of American history and their considerable historian's craft as well. Temple University's history department is also blessed with a competent and professional office staff. Pat Williams, Nancy Rymal, and Kathy Ewan from that staff have aided the completion of this work in many ways.

In and around Philadelphia several local history societies strive, against great obstacles, to resurrect the history of their communities. Diane Sadler of the Frankford Historical Society and Tom Pomager of the Bridesburg Historical Society deserve commendation for their effort.

As this study evolved into a publishable work several people contributed their considerable talents. Kara Raudenbush provided the photographs, and Roger Geyer drew the illustrations of the Maynard primers and lock and of percussion locks and caps. Pictures and illustrations are worth more than a thousand words; they are able to show clearly what words never could. Tony Costantino also deserves much credit: his mastery of the mysteries of computer technology and word processing programs made this study relatively painless, and Tony's lovely wife Eileen graciously accepted panicky phone calls for help at strange and inconvenient hours.

Last, but most of all, my wife, Rosemary, has made this study possible. Her love and support are never-failing.

Finally, this work is dedicated to the people of Bridesburg and Frankford and to the men and women who worked at the Frankford Arsenal both in this century and the last, for their perseverance, their dedication to their work, and their community spirit.

Introduction

As part of the surge of nationalism that followed the debacle of the War of 1812, the newly created Ordnance Department of the United States Army established several new arsenals to correct the disastrous shortages of arms, ammunition, and other military supplies that had plagued the army during the war. In 1816 one of these new arsenals opened in Frankford, Pennsylvania. Frankford was an established but small manufacturing community connected by Frankford Creek to the Delaware River just north of the city of Philadelphia. Although Frankford was within Philadelphia County, the site of the arsenal was sufficiently removed from the congestion of the city and far enough upriver to protect it from a seaborne attack, but it was close enough to the city to make the port and other local services accessible to the new arsenal. In the following year President James Madison dedicated the newly opened Frankford Arsenal.

Over the next fifty-three years the Frankford Arsenal evolved—from a small post at which soldiers of ordnance received, shipped, and maintained military supplies and manufactured small arms ammunition and other military equipment by hand, to a large-scale industrial complex in which a large civilian work force labored in the mechanized production of ammunition. How did this unanticipated transformation affect the officers and enlisted soldiers of ordnance of the United States Army? How did this change the relationship between the military and its civilian work force? How did it affect the way that each side perceived the changed relationship?

Like other Americans, historians have long been fascinated with science and technology. Late nineteenth- and early twentieth-century historians frequently have interpreted the rapid development of science and technology, as evidenced in machines and industry, as proof of America's march of progress. Some historians of technology have focused on changes in technology independently of external forces; they wax eloquent over sulfur, niter, and dynamite. Other historians have approached technology from an economic perspective; they usually explain the acceptance of a new technology or a new machine in terms of cost effectiveness or other economic considerations. Recently, a group of historians of technology (Merrit Roe Smith, Judith McGaw, Robert B. Gordon, among others) has adopted a different perspective. They argue that, in order to be understood, changes in technology must be studied in the contexts, social and cultural especially, in which they take place. They advocate the contextualization of the history.[1]

To approach historical clarity, I examine in several different contexts the developing technological changes that culminated in the creation of an industrial system for the manufacture of small-arms ammunition at the Frankford Arsenal.

I begin with a description of the village of Frankford as it evolved from a manufacturing to an industrial community and explain why the Ordnance Department of the United States Army selected this location as the site of one of its new arsenals. This chapter also emphasizes the differences between the village of Frankford and other early American communities, especially those in New England that have been taken as the prototype for American development. In describing the evolution from manufacturing to industry, I argue that the inhabitants of Frankford accepted this economic change and saw it as a continuity rather than as a divisive disruption.[2]

The focus then shifts from the community in which the Frankford Arsenal was located to the growth and development of the arsenal itself. This uneven and at times uncertain development occurred in four stages over more than fifty years. Until the Mexican war the arsenal remained a small army post that functioned mainly as a depot but also, sometimes, as a place where ammunition was manufactured and weapons refurbished. The first change followed that war when the Ordnance Department decided to enlarge the arsenal and transform it into a center for the mechanized production of small-arms ammunition. By 1854 a steam engine had been added, and steampower replaced handpower in mechanized production. The demands of the Civil War prompted the final two stages. The Ordnance Department doubled the size of the arsenal and erected several enormous buildings, each with a specialized industrial function. Then, to labor in this new industrial complex, the department employed a large civilian work force that both augmented and superseded the small detachment of enlisted soldiers of ordnance.[3]

In these three chapters that describe the early development of the Frankford Arsenal, several themes emerge. Since a number of factors (economics, war, politics) influenced the growth and development of the arsenal, alternating periods of stagnation and rapid development characterized its early history. Economic hard times retarded growth; prosperity promoted it. The depressions of 1819, 1837, and the mid-1850s and the inflation of the Civil War adversely affected the arsenal. Rapid growth followed wars or incidents that threatened military action. The periods of greatest development attended or followed the War of 1812, the bellicosity of Andrew Jackson's administration, the Mexican war, and especially the Civil War. Politics and political intrigue affected the arsenal both positively and negatively. The energetic administration of four especially capable secretaries of war (John C. Calhoun, Joel Poinsett, Lewis Cass, and Edwin McMasters Stanton) and several equally energetic chiefs of ordnance (George Talcott, Henry Knox Craig, and George D. Ramsay) promoted growth. The tenure of two former arsenal commanders, Craig and Ramsay, as chief of ordnance proved especially beneficial. But where politics promoted they also hindered, particularly during the bureaucratic and sectional infighting that immediately preceded the Civil War.

Since these three factors, economics, politics, and war, combined with each other and with other factors, the early history of the Frankford Arsenal was often chaotic, sometimes dynamic, but always uncertain. In this respect its history reflected the turbulent American experience in the first half of the nineteenth century, especially at those points where political events and economic change interacted.

Another theme that surfaces here is the changes in the United States Army and the public perception of it. Russell Weigley had shown that early in the nineteenth century Americans held their army in very low regard, and soldiers, officers especially, suffered from a sense of isolation. Ancient prejudices against standing armies in peacetime, republican egalitarianism, and the popular perception that the regular army performed poorly in the War of 1812 cemented public opinion against the military. By 1815 the War Department had decided that advances in European military technology had to be introduced into American military preparedness. This decision, along with the organizational and technological changes that it necessitated, began a long process of change in the United States Army; in the department itself; and in public attitudes toward both. The Ordnance Department became bureaucratic, scientific, managerial, and industrial. At the same time it also became much less isolated and more highly integrated into the mainstream of American public culture.[4]

However, this sense of isolation initially benefited the department. Being treated as pariahs fostered in the fourteen young officers of the Ordnance Corps a sense of brotherhood and camaraderie. They quickly developed a network that shared scientific ideas and technological innovations, almost as if they could use their expertise and accomplishments to justify themselves to a hostile American public. The coming of the Civil War and the subsequent transformation of the Ordnance Department (at least at the Frankford Arsenal) into a major employer dramatically and positively affected public opinion. A formerly undesirable neighbor became a valued and respected institution. Simultaneously, the sense of isolation experienced by both the officers and the enlisted soldiers of ordnance dissipated. When army officers evolved into employers, supervisors, and foremen, they might be trusted, respected, despised, or ridiculed — but they were no longer isolated. Through the Frankford Arsenal, the Ordnance Department became industrially, economically, and socially integrated into the community.[5]

Three other interrelated themes that involve historiographic argument weave through these chapters: the relative impact of the coming of industrialization on skilled workers, the perception and reaction of workers toward industrialization, and the comparative roles of private industry and the federal government in the promotion of American industrialization. Technology fundamentally changed the manufacture of small-arms ammunition. Originally, the combined efforts of differently skilled soldiers produced cartridges composed of paper, a lead bullet, thread, and gunpowder. More skilled soldiers measured and compounded gunpowder, a dangerous job, while others cut paper and thread or wrapped, tied, and packaged cartridges. The 1850 introduction of a hand-cranked machine and metallic caps changed the kind of work performed and the kinds of skills needed. Work became even more specialized, and with the adoption of steampower and the addition of more machinery it became faster and more complex.

Some historians have argued that industrialization denigrated formerly skilled artisans into unskilled factory workers. This argument might apply in some industries, notably textiles and shoes, but not in munitions production. Although the evolution of mechanized and industrial production at the Frankford Arsenal changed work, it did not necessarily deskill handicraft workers. In fact, as mechanized production became an industrial system it called forth new skills that had not previously existed, while old skills often retained their value.[6]

By the time of the Civil War the creation of this complex industrial system had evoked an equally complex work force. This group of several hundred workers varied in age, gender, skill, and therefore jobs performed and wages earned. How had these workers, their lives, and their families been affected by the creation of this industrial system? Is industrialization simply something that happens to workers, or is there a dynamic exchange, an interaction? How did these workers see industrialization?

Merrit Roe Smith has offered evidence to prove that the Ordnance Department originated industrial production in America at its armories at Springfield, Massachusetts, and Harpers Ferry, Virginia. However, in a recent dissertation Donald Hoke claims that after 1850 the Ordnance Department receded into stagnation and that private industry stepped to the forefront in industrial and technological innovation. Hoke's contention might hold true for the federal armories, but this study offers proof that in the first half of the nineteenth century the Ordnance Department consciously created an integrated industrial system of manufacture at the Frankford Arsenal well in advance of private industry. Indeed, I will cite several specific examples, including the Remington contract and the rifling machine, to demonstrate that as late as 1854 the technological capacity of the department exceeded that of private industry.[7]

In the end I return to Frankford to describe the complex, industrialized work force that labored at the arsenal in the years after the Civil War, and to the industrial community just beyond the gates in which these workers and their families lived. Using the models employed by Tamara Hareven in conjunction with information from the federal censuses of 1860 and 1870 and arsenal employment records, I concentrate on this work force as it encountered industrial labor, posing several further questions. Can workers use industrialization to promote individual and familial goals? Can relatively unskilled, lower-paid workers prosper in the industrial system? What is the boundary separating that which workers find tolerable and that which is unacceptable? By locating that boundary line, what can we discover about work and workers? And finally, can industrialization be a unifying force as well as a divisive one?[8]

This study, then, details the technological change in the fabrication of small-arms ammunition during the period from 1816 to 1870 as this process evolved from handicraft manufacture to systematic industrialized production. It also investigates the several contexts in which this evolution took place describing how and why the Ordnance Department both instigated this evolution and was affected by it. Finally, this study demonstrates the ways in which this technological evolution affected work, workers, and their communities.

1

The Village of Frankford, 1690–1850

Today, Frankford shares with other Philadelphia neighborhoods the burden of urban anonymity. It appears as a place-name on street maps or as a transit terminus for commuters. Historians have treated Frankford no more kindly. Yet Frankford has a long and important history that begins before the arrival of William Penn and continues to the present. During its first two centuries the village of Frankford experienced significant economic and social changes. Frankford and neighboring Germantown shared a number of important characteristics. Both were urban in their settlement, industrial and commercial in their economy, associational in their communal structure, and mobile and heterogeneous in their population.[1]

Local legend holds that Swedish settlers built a gristmill along Frankford Creek early in the second half of the seventeenth century.[2] While the date and the exact location of such a mill remain unverified, early records support the probability of its existence. On August 12, 1680, several Swedes asked the English court at Upland (Chester) to survey land along the creek to which they held warrants. One survey referred to the tract as "opposite land of Pelle Dolbos . . . land called mill creek." The survey also showed a road crossing Frankford Creek along the path of what became under English rule a king's highway, Frankford Road. Several other early surveys also mention both the mill creek and the road crossing it. With the coming of English rule, ownership of the land in Frankford began to change hands rapidly. In succession it passed from the Society of Free Traders to the Frankfort Company to, by 1687, four English Friends, Thomas Fairman, Robert Adams, Henry Waddy, and Jonathan Harper. Briefly, while in the possession of the Society of Free Traders, the area south of Frankford Creek was identified as the Manor of Frank. Presumably, the name Frankford comes from the identification of the area either as the Manor of Frank or with the Frankfort Company. Already by 1698 most of these grants had been subdivided or had changed hands entirely. Indeed, this rapid and continual exchange of land by sale, lease, or rent remained one of the constants in early Frankford history. In the same year another English Friend, Thomas Parson, operated a gristmill on the creek, adjacent to the Frankford Road.[3]

During the eighteenth century the village of Frankford continued to grow as a manufacturing community. The original mill changed hands once more. Joseph I. Miller purchased it along with fifty acres of land and added a sawmill to the gristmill. During the 1750s a number of Germans began to settle in Frankford. Two brothers, Jacob and Rudolph Neff, bought land and opened a wheelwright's shop. Before the end of the century they owned several dwellings and lots in the village, and they either operated a mill or at least owned property along the millrace above Miller's property. In 1775 Oswald Eve erected a gunpowder mill less than a mile up Frankford Creek on the northwestern edge of the village. Eve's ownership proved to be brief and stormy. In 1778 the Pennsylvania Supreme Executive Council convicted Eve of being a Tory and confiscated the powder mill. It subsequently passed into the hands of John Decatur, father of naval hero Stephen Decatur, and then in 1806 to a local war hero, Gen. Isaac Worrell.[4]

The manufactories of Frankford were not confined to the banks of Frankford Creek. Smaller handicraft industries and commercial establishments not dependent on waterpower stretched northward along Frankford Road (also referred to as Frankford Street, the King's Highway, and later Main Street) as it rose from its junction with Frankford Creek. Shops and dwellings of carpenters, joiners, millwrights, weavers, smiths, coopers, masons, cordwainers, coachmakers, a clockmaker, a silversmith, and a bookbinder clustered together on both sides of the road in less than a mile. The commercial district terminated in the tanning

FIG. 1. This building sits on Frankford Creek at the site of Oswald Eve's gunpowder mill—one of the oldest industries in the village of Frankford, dating from 1775. To the right and the left in back are nineteenth- and twentieth-century additions. These additions expanded the site into an elaborate industrial complex, complete with its own steam plant.

yard of Nathan Harper, descendant of Jonathan Harper, one of Penn's first purchasers.[5]

By the end of the eighteenth century the urban manufacturing village of Frankford contained nearly one thousand inhabitants, occupying just more than one hundred dwellings. This concentration compressed about one-half of the entire population of the township into an area approximately one mile square. These mills, shops, and homes were constructed along the two major economic axes of the community, Frankford Creek (and Adams Road paralleling it) and Frankford Road. Together they served as the lifelines of the village and delineated its economic existence. Travelers to and from Philadelphia stayed in one of several inns. Farmers from the surrounding countryside had their grain ground, bargained for sawed lumber, and purchased other necessities or luxuries there.[6]

The homes of the people of Frankford gave evidence of considerable economic diversity. Types of structures and their assessed values ranged from the $150, one-story log cabin occupied by Jonathan Dungan, a cooper, to Robert Waln's $6,000 mansion. Although several prosperous men and wealthy widows owned a number of leased properties, 40 of the village's 202 heads of household owned the dwellings that they occupied. These included the houses of Jacob Harper, wheelwright, valued at $200; Andrew Schoch, tailor, $500; Thomas Gillingham, carpenter, $600; and Benjamin Love, also a wheelwright, $2,000.[7]

As the nineteenth century began the village shared the economic optimism that marked the whole Delaware Valley as the young American republic rebounded from the postwar depression. By 1810 Frankford presented a complex image of growth and mobility along with social and economic stability. The total population of the borough had increased to 1,233 people.[8] The old mill had changed hands once again; two wealthy local landowners, Abraham Duffield and his brother, Thomas W. Duffield, Sr., purchased it in 1800. But many of the same names reappear in the county tax lists and census records.

The fortunes of Andrew Schoch, the tailor, seem to have declined. Even though his household still included seven inhabitants, and he owned a small lot and frame dwelling, its assessed value had declined to $50. Jacob Harper, the wheelwright, remained stable: he still owned a half-acre lot and a frame dwelling valued at $180. But Benjamin Love and Thomas Gillingham, both prominent Friends, had prospered. Gillingham now described himself as a blacksmith, and he owned five acres, another small lot, a stone house, and a frame dwelling as well as his smith shop. Gillingham's household was also large at ten people.[9] Benjamin Love listed his occupation as lumber merchant. He owned a large number of different properties and dwellings, including nine and a half acres, two lots, two stone dwellings, two frame dwellings, a workshop, a granery, and a lumber house. These cases demonstrate the complexity of the changes that affected this and other American communities at the start of the nineteenth century. They also reveal the degree to which the people of Frankford valued property and its acquisition.[10]

From 1790 to 1850 some family names—Worrell, Shallcross, Castor, Foulkrod, Deal, and others—appear repeatedly in the records, indicating a high level of continuity and stability in the community. Typically, these families lived in the village and were identified with manufacturing and commercial occupations, and they held both town lots and substantial holdings in the countryside. Such an investment strategy probably enabled them to prosper in times of growth and protected them from economic adversity.

Rural land and town lots changed hands in their entirety or in subdivided parcels. Occupations changed; new ones appeared, but old ones remained. This constant flux defies facile analysis, but it clearly represents something more than simple mobility. In Frankford rapid exchange of land and property occurred. Small entrepreneurs, great families, and wealthy individuals came and went. All of this movement revolved around individuals trying to prosper. Whether few, some, or many succeeded seems not to have mattered to the other residents of Frankford bent on economic advancement.

The records portray early nineteenth-century Frankford as a place of contradictions, of persistent stability as well as seething change. Only the degree to which both of these tendencies coexisted distinguishes Frankford from other early nineteenth-century American communities.

During the period from 1812 to the middle of the next decade, however, three events tore away the veneer of stability. From its origin Frankford had been a manufacturing community, but now true industrialization appeared. In 1809 Samuel Martin, an English immigrant, opened the first textile mill, an early portent of Frankford's future. In 1816 Isaac English started a pottery. By the following year two umbrella frame factories were in operation, and a third opened in 1830. Before 1817 Richard and Danforth Woolworth established a jewelry shop along Frankford Road. In this shop Mathias Baldwin, Philadelphia's preeminent nineteenth-century industrialist, received his initial training in metallurgy.[11]

In 1813 yet another textile mill established by an English immigrant, Henry Whitaker, began operation along Frankford Creek less than a mile north of the borough. In 1835 Whitaker Mills became the first American factory to convert to steampower. The engine was locally manufactured by Alfred Jenks of Bridesburg. This factory remained in production through the nineteenth century, owned and operated by the same family.

It is difficult to tell at what point Frankford ceased to be a manufacturing village and became an industrial community, for hand manufacturing persisted throughout the nineteenth century; but Whitaker Mills represents a significant step in the transition. Henry Whitaker owned the property, the mill buildings, and the tools and equipment. Whitaker's account books show clear evidence of wage labor. Workers were paid by the day; they used no tools of their own, and there is no evidence of any master-journeyman-apprentice relationship.[12]

In 1816 one more industrial site was developed at the opposite end of Frankford Creek, where it joins the Delaware River. At the end of the War of 1812 the ordnance branch of the United States Army had begun inspecting land near the village, along the creek, for a site for an arsenal under its own control. In 1815 the Ordnance Department took up an option to buy a parcel of land identified as the Prentiss plot, but then the department decided to purchase another, more expensive property near the junction of Frankford Creek and the Delaware River. On May 27, 1816, the department, represented by Capt. Joseph H. Rees, purchased from Frederick Fraley and his wife, Catherine, over twenty-two acres on the north bank of the creek about four hundred yards from its confluence with the river. The purchase price was $7,680.75. The Ordnance Department considered this the official beginning of the Frankford Arsenal.[13]

This property had a long and murky history before it came into the possession of the federal government. Local legend held that it was the site of a Leni-Lenape campground, although the painstaking 1979 archaeological survey done by John Milnor Associates found no evidence of aboriginal encampment. During the first forty years of the eighteenth century the property, like other real estate in the area, changed hands several times. By 1742 Andrew Hamilton owned a great estate of 282 acres whose southern boundary followed the creek. In 1799, in recognition of its commercial importance, the Pennsylvania assembly designated the creek as a public highway. Finally in 1810 Frederick Fraley, an immigrant Swiss baker, purchased the property.[14]

Several sources that describe the early history of the Frankford Arsenal refer to the site as a "cantonment" on which a "sham battle" called "Point-no-Point" took place on June 22, 1808. Thomas Guernsey's "History of Frankford" seems to be the common source for these references. However, Guernsey cites no authority for his information, and the 1979 archaeology survey found no evidence of either a battle or a prior military encampment. One of Fraley's descendants suggests that the term "cantonment" stemmed from Fraley's Swiss heritage rather than from a prior history as a military camp.[15]

A wharf, built with materials from an excavation for a cellar, was the first major construction project at the arsenal. Upon its completion boats landed loads of stone to be used by the enlisted laborers, soldiers of ordnance, to erect a storehouse and temporary quarters for a command that consisted of two officers and twenty-two enlisted men. In 1817 President James Madison presided at the laying of the cornerstone, and a productive future seemed assured. The Ordnance Department had originally planned to spend $35,000 on construction, but in 1817 Congress, freed from the demands of war, failed to pass the necessary appropriations. Construction came to a standstill, and the Frankford Arsenal languished.[16]

The conclusion of the War of 1812, followed by the Panic of 1819, affected more than just the initial construction of the arsenal. Other local industries disappeared entirely. Samuel Martin's textile mill prospered briefly on "a large government contract," but a fire destroyed the mill after the war. Martin did not rebuild the mill, and he returned to England shortly thereafter. However, he left

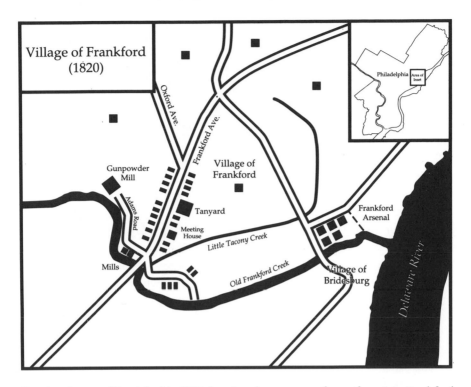

Fig. 2. A map of Frankford in 1820 showing the two axes of manufacturing: Frankford Road and Frankford Creek.

behind five sons who in the 1830s and 1840s contributed to the resurgence of the textile industry in the Philadelphia area. One of the three umbrella factories also closed, perhaps driven to the wall by bitter competition. Many craftsmen and manufacturers who had briefly appeared in earlier censuses and tax lists were now described simply as "gone." Among others, the panic claimed Robert Waln, one of the great merchants of Philadelphia and the wealthiest property owner in Frankford. By 1820 these calamities had substantially altered the economic topography of the village of Frankford.[17]

During the early 1820s another dire event, religious turmoil, threatened the social fabric of this industrializing village. Although some of William Penn's first purchasers had founded Frankford, diversity marked the religious life of the village throughout the eighteenth and early nineteenth centuries. The Keithian controversy had resulted in the establishment of the Oxford Trinity Episcopal church in the far western corner of the township. In 1775 the Presbyterian church founded a congregation in Frankford. However, German immigrants transformed it into a German Reformed church, and during the revolutionary war the Pennsylvania Supreme Executive Council incarcerated Hessian prisoners there. In 1809 a Baptist church joined the list of houses of worship. And at some point the substantial free black community formed an African Methodist Episcopal church.[18] Yet even with this proliferation of churches, most inhabitants of Frankford remained unchurched, or at least their names do not appear on existing church lists.[19]

Even with religious diversity, however, a Quaker elite dominated Frankford. Most of the leading landowners, craftsmen, and manufacturers, as well as officers of the borough government, belonged to the Religious Society of Friends. But in 1827 another schism, the Hicksite controversy, ended the Quaker domination of Frankford. In two years, 1827 and 1828, the Frankford Monthly Meeting "dismissed" at least forty-two members of the society. This wholesale dismissal resulted in a volatile and embarrassing situation. Since the expelled constituted the majority of the membership of the Meeting for Worship, they remained in possession of the meetinghouse, the Taconic Meeting. The remaining minority purchased land and built a new meetinghouse, the Orthodox Meeting.

Robert Doherty argues that in places where the Hicksites formed a majority, they clung to a sectarian past and stood as a conservative rearguard who resisted the encroachment of industrialization and other changes. The case of Frankford does not verify Doherty's contention. In Frankford some of the dismissed Hicksites such as Nathan Harper, proprietor of one of the tanyards, represented older, established, and perhaps threatened crafts, but others such as the Gillinghams had prospered with the economic transition that accompanied industrialization. The case of the Gillingham family, among the oldest and most prosperous in the community, is especially baffling. After having been dismissed, three Gillingham brothers and their families left Frankford entirely and resettled in Nottingham Creek, Maryland. In other families some members remained in communion while others were dismissed. By 1828 the Hicksite schism had shattered the Quaker elite of Frankford. In its wake remained a vacuum awaiting the emergence of a new elite in an industrially oriented community.[20]

While the Friends suffered, other congregations benefited. In 1820 Charles Grandison Finney's revival, the Second Great Awakening, swept through Philadelphia, and its effects reverberated in Frankford. Until 1820 the membership of the Frankford Presbyterian church remained static. From 1800 to 1820 the congregation admitted only fifty new members. Eighteen new souls "came to grace" in 1820, eleven in 1821, five in 1822, eight in 1823, fourteen in 1824, and eleven again in 1825. The pattern of admitting new members followed that

already noted by Mary Ryan: older female family members joined first, followed
by either younger females or related males.[21]

As a result of this turmoil and change, during the 1820s the Quaker elite lost
its dominance, the Presbyterian congregation swelled, and a number of new
churches joined the older ones, thereby enhancing the already marked pluralism
of Frankford. Both the economy and the religious and ethnic pluralism of this
urban industrial community grew. With the revival of prosperity following the
Panic of 1819, new industries and workers poured into Frankford.

In spite of the social and economic setbacks between 1817 and 1820, the
industrial foundation of Frankford had remained intact. Of 228 households
polled in the 1820 census, the majority, 153, identified their occupation as either
manufacture or commerce—but with a nativist flavor, because the same census
listed only six aliens out of a total population of 1,405. The new industrialization
of the 1820s rapidly diluted that nativist tint in the population of Frankford.

Not surprisingly, much of the industrial development that took place during
the 1820s and afterward centered on the production and finishing of textiles. In
1821 Samuel Pilling erected a calico factory, although records do not indicate
whether calico was woven or simply printed there. Within several years two
English immigrant entrepreneurs, Jeremiah Horrocks and John Large, opened
factories for the bleaching, dying, and finishing of cloth. Horrocks may have
been the first Frankford industrialist to employ black laborers, although records
suggest that a black worker had earlier labored at Worrell's powder mill. In 1827
John Garsed and his brother operated the Frankford Woolen Mill. Upon its
completion this mill, which still stands, must have dominated the Frankford
skyline. Very much resembling the mills of Lowell, it stands four stories high,
thirty feet wide, and ninety feet long, bordering on the millrace just below the
site of Worrell's mill complex. Even though the textile industry dominated this
spate of development, other industries were represented, including Christopher
Wesener's Frankford Chemical Works, constructed in 1829. By the end of the
1820s Frankford's long evolution toward industrialization approached a significant turning point, for along with these new industries came a new immigrant
work force.[22]

Since immigrant English industrialists such as Henry Whitaker, John Large,
and Jeremiah Horrocks had begun almost all of these new factories, the arrival of
a large number of immigrant workers comes as less than a surprise. Frankford's
nativist hue changed quickly. New immigrants poured in from England, Ireland,
and Scotland, and they brought with them new occupations such as bleacher,
dyer, starchmaker, and calico printer to enhance Frankford's already considerable
list of industrial occupations. Even before 1830 Frankford had acquired an ethnic

Fig. 3. This building, perhaps fifty yards downstream from Oswald Eve's gunpowder mill, housed the Frankford Woolen Company. Constructed in the classic long, narrow, three-story layout favored by the early nineteenth-century American textile industry, it dates from perhaps the 1850s; but the cornerstone of a smaller, attached building is dated 1821.

neighborhood, one of the main features of nineteenth-century industrialization in the Philadelphia region. "Little Britain" consisted of a cluster of immigrant dwellings in the older southwestern part of the community. Ironically, this ethnic neighborhood sat at the foot of the hill where wealthy industrialists and professionals began to construct imposing Victorian homes, and it was juxtaposed to the poorer black community on the opposite side of Main Street in the northeast quadrant of the village. Thus, before midcentury Frankford started to experience the kind of polarization that marked metropolitan Philadelphia later in the century.[23]

In a recent study on the mills of Manayunk, Cynthia Shelton has identified a work force of British immigrants similar to those who began to appear in Frankford in the 1820s. In an extended discussion of the strikes and other job actions in which the mill workers of Manayunk engaged, Shelton concludes that the workers focused on specific economic objectives while they eschewed broader social ones. Similarly, Mary McConaghy argues that the only identifiable strike in Frankford, the 1842 strike at Whitaker Mills, had specific economic goals, a restoration of lost wages.[24]

The workers, both native-born and immigrant, at Whitaker Mills in Frankford and in Manayunk shared the same view of industrialization. Both groups saw it as an acceptable economic condition. Native-born Americans saw it as a vehicle for the individual acquisitiveness that marked Frankford from its origin. Immigrant workers had no illusions about the new industrial technology. They saw factory work clearly and realistically as backbreaking labor, but labor that held out the hope of a comfortable old age and a better life for their children. This hope proved to be an illusion for many, but that fact in no way diminished its appeal to an early nineteenth-century immigrant worker. If any craftsman in Frankford mourned the lost world of artisanal independence, his voice was lost among the clatter of machinery and the bustle of Main Street.[25]

Before the middle of the nineteenth century, Frankford had evolved into an industrial community in all of its essential features. Despite the persistence of some small and handicraft manufacture, industrial production formed the basis of Frankford's economy.

The community reflected this industrial foundation. Indeed, industrialization was the unifying factor in this heterogeneous society made up of rich mill owners; an emerging white-collar, professional, and commercial middle class; and industrial workers and marked by ethnic, religious, and cultural diversity.

By 1850 the population of Frankford exceeded five thousand, and Frankford became a victim of its own growth and prosperity. Heretofore separate communities merged and boundaries disappeared. In 1854 the Pennsylvania legislature lumped Frankford and other previously independent communities into the city of Philadelphia. Ignominiously, Frankford became merely the Twenty-Third Ward. Today Frankford shares the fate of older industrial areas in the northeastern United States. Each day commuters pass over it on their way to more affluent suburbs, and empty factories stand as melancholy reminders of a nobler past.

2

The Early Years of the Frankford Arsenal, 1817–1830

Since, as so often happens, historical documents fail us, it is left to our imaginations to picture the opening of the Frankford Arsenal. In May 1817 with celebratory republican ceremony, former president James Madison presided at the laying of the arsenal's cornerstone. Our imaginary crowd, the small detachment of soldiers of ordnance in their military uniforms mingled with the local citizenry, stood at the foot of the steps of the West Storehouse, the arsenal's most imposing building. To the left the commanding officer's quarters marked one corner of the parade ground, while the buildings that housed the rest of the post's soldiers were on the right. To the rear about a hundred yards away was the arsenal's workplace. Just behind the West Storehouse a wooden fence and a rutted, unpaved road barely separated the arsenal from its rural environs. Even before the sounds of the speeches died, the arsenal fell on hard times when, for economic reasons, Congress subordinated the briefly independent Ordnance

Department to the Artillery Branch of the army. As would happen frequently in the arsenal's history, building and expansion came to a grinding halt. In more than one instance bureaucratic lethargy and congressional parsimony defeated farsighted planning. For the next fifteen years the arsenal remained small and insignificant. Along with its parent, the United States Army, it carried the twin onus of early American antimilitarism and early nineteenth-century egalitarianism. In addition to these two burdens, local animosity also plagued the Frankford Arsenal during its pre–Civil War days.

U. S. Arsenal, near Frankford, Pa.

FIG. 4. A woodcut that appeared in the *Saturday Evening Post* on April 7, 1832. It emphasizes the bucolic appearance of the Frankford Arsenal in its early years. A post-and-rail fence and a muddy road border the arsenal grounds—a sharp contrast to the stone wall and the iron fence that would be put up later.

Even though three important and permanent buildings were added in 1819 and 1822, for the first twenty years the arsenal struggled against obscurity. The career of Joseph H. Rees, the first commander of the post, highlighted this impression. In March 1819 Captain Rees received a letter from the ordnance

FIG. 5.   Building 11, the original arsenal, later used to house administrative offices.

office in Washington, D.C., that indicated problems with his financial accounts: "As the matter now appears the impression is very unfavorable." Less than two months later he received another more imperative missive that mentioned the sum of $8,000 and added, "your continuance in the Service will badly depend on it." Rees was then summoned to the ordnance office. Rees apparently settled the issue satisfactorily, for he remained in command. The following year, however, he seems to have suffered an illness that prevented him from exercising command. A junior officer, Lt. Martin Thomas, Jr., signed letters and conducted business usually reserved to the commanding officer. In February 1821 Lieutenant Thomas wrote to a Mr. Lee in the adjutant general's office notifying him of the death of Captain Rees on February 6. Six different commanders, most of whom were junior officers, succeeded Captain Rees over the next eleven years.[1] (See the Appendix for a complete list of commanders of the Frankford Arsenal from 1816 to 1876.)

This early instability and apparent unimportance are deceptive, however, for

FIG. 6. Building 18, the original storehouse (also referred to as the West Storehouse), dedicated by President James Madison in May 1817.

during this period several major themes and patterns were established that not only influenced the growth of the Frankford Arsenal as it evolved from primarily a storage depot to a place of manufacture, but also mirrored the development of the American arms industry and its relationship with the federal government. By examining these earliest decades of the arsenal, we can see the early, groping attempts of the United States Army to promote uniformity in the manufacture of arms and ammunition, especially by increased use of inspection and gauging. This examination also provides a picture of its early work force that stands in sharp contrast to that of its post–Civil War industrial workers. Moreover, a study of the early period allows a description of the contract system, an initial attempt by the Ordnance Department to assure both quality and quantity of military supplies. The flaws in that system ultimately led the department toward the industrial system of munitions production at Frankford.

Fig. 7.   Building 1, the commanding officer's quarters, built in 1821 and renovated several times in the late nineteenth century, it served this purpose until the arsenal closed.

THE CONTRACT SYSTEM

In the early nineteenth century a strange dualism marked the relationship between the American arms industry and the federal government. The government manufactured, in its own armories at Springfield, Massachusetts, and Harpers Ferry, Virginia, most of the small arms,[2] rifles, and muskets it required, but it also purchased from private manufacturers a large variety of arms, ammunition, and other war material. This material was procured through a unique and complicated contract system. From the early nineteenth century through the Civil War the War Department purchased vast amounts of goods by this method. Under this system the Ordnance Department gave a contractor an order for a large number of items, such as 2,500 muskets, to be filled over a period of time (as long as five years) and at a set price. The government provided a pattern musket that the manufacturer used as a guide. The government also sent to the manufacturer's shop an inspector paid by the Ordnance Department and equipped with sets of gauges to examine each piece as well as the assembled, final product. The inspector then stamped each piece that passed his inspection. When the manufacturer finished a specific number of items they could be delivered to the designated arsenal or armory for payment after a final inspection. Both parties benefited. The manufacturer, blessed with a long-term contract for a large number of weapons, was guaranteed a profit even if the initial production required extensive retooling. Also, if the manufacturer justified any additional costs the Ordnance Department might increase the price paid. This system appeared to guarantee the Ordnance Department uniformity of manufacture. Munitions could be made to fit any musket or rifle of the same model, no matter who was the manufacturer; and replacement parts could be furnished to repair weapons, which tended to be treated very badly in the field, admonitions, threats, and punishments notwithstanding.[3]

The contract between Marine T. Wickham and the Ordnance Department is a good example of the contract system. In 1822 Wickham, one of a fairly large number of established Philadelphia arms manufacturers, received a thirty-month contract to deliver to the Frankford Arsenal three thousand stands of arms in parcels of three hundred.[4] This thirty-month term was unusually short, since contracts were generally for five years. For each stand that passed a final inspection by the arsenal commander or his designated subordinate, the manufacturer received twelve dollars.[5]

The contract system appeared to be not only an ideal system for both the Ordnance Department and its contractors, but also an important step in the evolution of American technology beyond the arms industry. Since the system seemed to guarantee a profit for the manufacturers, cooperation among them and between the industry and the national armories developed, and companies quickly exchanged innovative ideas and technology. By contrast with today's climate of industrial competition and closely guarded secrets, this sharing of ideas and technology is difficult to explain. Perhaps a long-dead spirit of mechanical camaraderie or the shared lust for government payment explains its existence. The system of patterns, gauges, and inspections of each piece promoted interchangeability of parts, precision manufacture, and its rapid subdivision. This, in turn, both encouraged mechanization and had an impact on the role of the worker, who was now faced with increased specialization and subdivision of his job.[6]

Letting contracts for a large variety of goods—from muskets and gunpowder to rifle slings and bayonet scabbards—receiving, inspecting, temporarily storing the finished goods, and finally redistributing the material upon order from the Ordnance Department constituted the main tasks of the Frankford Arsenal. This system enabled the arsenal to procure for the American military the necessities of war from both far and near.[7]

The contract system provided the War Department with hugh quantities of a large variety of manufactured goods and raw materials. However, several problems plagued the system. Frequently, contractors failed to meet deadlines, tried to pass off inferior equipment, or reneged on contracts. And these difficulties seemed to occur at times of the greatest need. Exasperated, cajoling, and sometimes angry letters passed between various arsenal commanders and contractors. In times of economic instability (frequent in the first half of the nineteenth century), some suppliers found long-range government contracts a losing proposition because the instability of currency made set prices meaningless. Others, disillusioned by slow payment, refused to renew contracts. Since the system depended on the honesty and integrity of the government inspectors, human frailty bred corruption. Later, especially during the Civil War, contractors with powerful political connections tried to force the Ordnance Department to receive shoddy goods. Even with endless systems of gauges, patterns, inspections, and reinspections, the contract system fell victim to one particularly fatal flaw: lack of uniformity and standardization in production.

In July 1834 the master armorer of the Frankford Arsenal, Emerson Foot, was in the process of examining, for the purpose of repairing and refurbishing, 145 cases of muskets from the Bellona Arsenal near Richmond, Virginia. Even though the guns had all been manufactured from pattern models at one of the national armories, the length of the barrel varied from thirty-eight to forty-five inches and the size of the caliber by as much as two-hundredths of an inch. Weapons produced by private manufacturers often displayed greater variations in tolerances.

As nineteenth-century warfare became more deadly and more technologically advanced, manufacturing inefficiences such as these became less and less acceptable to the Ordnance Department. These flaws in the contract system forced the department to take two tremendous steps fraught with grave implications for itself, its arsenals and armories, and the future of American industry. First, it became more self-reliant, producing more and more of its own weapons and munitions and thus guaranteeing the growth of at least some of its installations. Second, hoping to promote product uniformity and cost and time efficiency, it came to foster technological innovation by using machines to replace hand manufacturing; by using improved machines to replace older, less efficient ones;

or by making increasingly detailed inspections with more elaborate and accurate gauges. These changes promoted more rapid industrialization and helped to redefine the meaning of work for both skilled and unskilled workers.[8]

Most of the contractors with whom the Frankford Arsenal conducted business appeared for a brief time in the arsenal's records and then disappeared. But one company, E. I. duPont de Nemours, became the major partner of the arsenal— and evolved into one of the dominant corporations in industrial America. Some sources imply that the arsenal was founded where it was due to its closeness to the duPont powder mills, located on the Brandywine River just above Wilmington, Delaware, and within easy shipping distance via the Delaware River to the arsenal wharf. This was clearly not the case. The duPont mills were not the only powder mills in the Delaware Valley region. One, in fact, operated on Frankford Creek within two miles of the newly founded arsenal.[9]

In 1801 John Decatur owned this extensive operation, which consisted of three stamping mills and a graining mill. In the same year E. I. duPont visited the mill and considered purchasing it. However, a Pennsylvania law that forbade foreigners from owning property prevented him from doing so. In 1806 Gen. Isaac Worrell, a local revolutionary war hero, bought the powder manufactory. As often happened at early gunpowder mills, explosions damaged Worrell Mills in 1810 and again in 1811. By the War of 1812 the owners had repaired the mills and solicited orders through their representative in Congress. The sketchy early records of the arsenal and the Ordnance Department offer no convincing proof of purchases from Worrell Mills, but if, as records indicated, the Ordnance Department carefully surveyed the industrial capacity of the area prior to purchasing the Fraley tract as the site for the arsenal, it must have been aware of the existence of the Worrell powder mills and taken its presence into consideration.[10]

In spite of the fact that a number of influential people in the federal government knew E. I. duPont, including President Thomas Jefferson, there exists stronger proof that in the eyes of the newly created Ordnance Department the duPont powder mills were relatively unknown. In the early fall of 1819 Captain Rees received a letter from Col. Decius Wadsworth asking the captain to investigate the duPont family and the value of their property, which the duPonts wanted to use as surety for a contract with the Ordnance Department. After duly conducting such an inquiry Captain Rees responded laconically that the duPonts "had given evidence of extensive real estate holdings." Only then did the chief of ordnance decide, "I should prefer the powder to be received from Messrs. Du pont." On July 11, 1822, Lt. Martin Thomas, now in temporary command, notified Col. George Bomford, the new chief of ordnance, that bins were being built to hold 105,000 pounds of niter. The December 27, 1825, records report "by sloop from Wilmington 25,941 lb. of saltpeter from E.I. DuPont & Co." Thereafter the name E. I. duPont appeared in the records of the Frankford Arsenal with increasing frequency, either in reports of shipments received or concerning a trip of the commander to Wilmington to test batches of powder. Even though (as we shall see in subsequent chapters) the duPont Company did not emerge as the dominant supplier of musket, rifle, and cannon powder to the Ordnance Department until the Civil War, by the mid-1820s it held a position of great importance in the complex contract system.[11]

Along with large quantities of gunpowder or its component elements being unloaded at the arsenal wharf, the acting storekeeper,[12] usually a junior officer, also received large shipments of lead. For example, a letter of October 16, 1821, notes the docking of a sloop with a cargo of 32,360 pounds of lead. In addition to lead and gunpowder, the returns of quarterly expenditures listed payments for cartridge boxes, "2300 cartridges each 92 cts. . . . $2116.00," paper, and bullet molds. Obviously, then, besides serving as a place of deposit, refurbishment, and distribution, the laborers at the Frankford Arsenal produced large quantities of musket and rifle cartridges. Evidence of shipments of these cartridges confirms this conclusion. On November 13, 1820, Lieutenant Thomas wrote to Capt. William C. Beard, the commander of the Seventh Infantry Regiment, advising him of a shipment of "1000 ball cartridges."[13]

These cartridges represented a significant improvement in munitions technology. As late as the revolutionary war each soldier often produced his own bullets from lead issued to him. Gunpowder was also issued in bulk. Bullets were almost never uniform in size and were just as likely to roll out of the barrel, or to jam the barrel and ruin the musket or rifle, as they were to fire correctly. Similarly, powder poured into the muzzle of the barrel in inexact amounts was as likely to blow up the weapon or to propel the bullet weakly as it was to fire accurately.

By the end of the eighteenth century paper cartridges that contained both the bullet and a premeasured amount of powder had come into general use. In the 1820s laborers at the arsenal produced these cartridges. Since this production constituted the most important work done there in terms of the future of the installation a close examination is warranted.

The process began with the preparation of the musket ball. Originally, the balls had been cast from molten lead, but by the 1850s they were pressed by machine. A worker took the pressed ball, laid it on a piece of heavy brown paper that had previously been cut into "trapezoids 4.33 in. by 5.25 in. by 3 in.," and rolled it into a cylinder formed about the ball. Next, the worker tied the paper cylinder at one end with a thread in two half-hitch knots and placed it in a tray with the untied end facing up. The trays were taken to the charging room, where

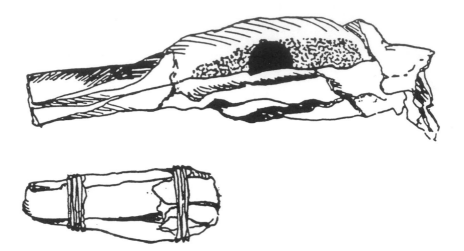

Fig. 8. The paper cartridge manufactured at the Frankford Arsenal. At bottom is the completed, wrapped cartridge, at top is the unwrapped cartridge, showing the bullet and the charge of gunpowder.

the cartridges would be filled with powder. The size of the funnel varied according to the number of grains of powder being inserted; the number ranged from 30 to 110 grains. Once filled, the cartridges were bundled into packages of ten and boxes of one thousand. The boxes, of a specific material and dimensions, were labeled and painted olive. Each step in this process required special "utensils" and handling according to painstakingly detailed instructions. Notice also that the production demanded a number of skills—judgment and eye-hand coordination, among others.[14]

Similarly, ordnance regulations delineated the number of workers to be employed, their exact functions, and the expected quantities to be produced in specific time periods. Thirty to thirty-five thousand musket balls could be made in eleven or twelve hours. Cutting the heavy paper required one cutter and one assistant. Making the cylinders required "1 Master; 10 men to roll cylinders; 1 to fill them, 4 to fold, 4 to bundle. . . . Boys or girls from 12 to 18 years may be advantageously employed." And "10,000 musket cartridges are made and bundled in 10 hours. . . . "[15]

By 1830, however, munitions technology was poised on the edge of several important changes that would have an impact on warfare in general and the role of the Frankford Arsenal in particular. As early as 1823 arsenal records referred to Joshua Shaw's testing of "a piece of ordnance you have invented." Shaw was working on a new method for firing a musket while other American inventors labored on improving ammunition.[16]

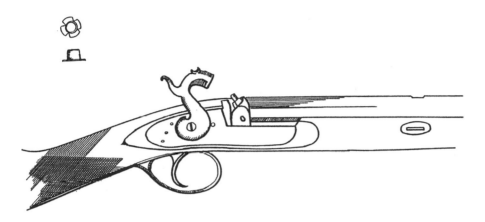

FIG. 9. The percussion lock. Above is the percussion cap. To fire the musket a soldier placed the inverted cap on the cone under the hammer and then pulled the trigger.

In addition to rifle and musket cartridges, the skilled laborers at the Frankford Arsenal also produced a number of other items ranging from individual artifacts such as "brushes and picks" ("this month we fabricated 1000 brushes and picks") to belts for infantry and artillery men that included complete sets of tools and other equipment ("with the force now employed we can complete about 240 sets of belts, per week"), as well as more specific instruments and gauges ("to have 5 sets of instruments made for the purpose of inspecting shells").[17] Two lists provide a better idea of the extensive variety of items manufactured at the post. In December 1819 Colonel Wadsworth ordered Captain Rees to have the following

prepared by March . . .
15,000 Priming tubes[18]
100 gunners belts with implements complete
1000 P(?) . . .

100 Haversacks for cannon cartridges
30 gunners quadrants.

In an inventory sent to Colonel Bomford in December 1832 the post commander listed the following items, some made at the arsenal and others some products of the contract system, on hand:

Cartridge boxes 2
Cartridge belts 953
Bayn't belts (private) 283
    "   "   (sgt.) 250
    "   "   (old pattern) 493
gun slings 43
Bay't Scabbards 1908
brushes and picks (white) 2041
Belt plates (yellow) 2492
    "   "   (white) 727
Rifle Flasks 4173
    "   sets of accouterments 1377
Pistols Rifle caliber 1438
    "   Musket 199[19]

Besides making an impressive variety and quantity of things and acting as a depot for the contract system, the Frankford Arsenal and its workers also performed a third function: they repaired everything from the largest artillery piece to the smallest measuring instrument and pattern item.[20]

The contract system appeared to fulfill the Ordnance Department's need for arms, ammunition and other military supplies, but later the demand for greater technological sophistication, especially in wartime, revealed glaring flaws in the system.

THE EARLY WORK FORCE
OF THE FRANKFORD ARSENAL

Even though arsenal records use terms such as "fabricated" and "manufactured," it should be emphasized that all of the work at the Frankford Arsenal was still handicraft labor done mostly by skilled craftsmen. Most of the tools identified in receipts and inventories were simple hand tools such as files and vises. By the 1820s water-powered machinery was being used in several American industries including the national armories at Springfield and Harpers Ferry. However, the machine age had not yet reached the Frankford Arsenal. Even though other industries along Frankford Creek above the arsenal used waterpower extensively, none was used at the arsenal.

From the opening of the arsenal until the middle of the 1850s the work force consisted of a relatively small detachment of enlisted men, occasionally augmented by hired civilian laborers. The number usually hovered around twenty-five, although it dropped as low as fifteen during the depression following the Panic of 1837 and rose as high as forty-five just after the Mexican war. The size of this group of working men, however, belies its significance, for beginning in the 1830s and continuing in the 1840s, 1850s, and through the Civil War the workers at the Frankford Arsenal experienced the coming of the new industrial age in advance of most other American workers. By the end of the Civil War complex changes transformed these handicraft artisans into industrial workers, toiling in a brand new workplace and laboring with hundreds of others to the rhythm of powerful machines. Yet the results of this cataclysmic upheaval were far from uniform. Some workers became clerks and supervisors, masters of the arcane skills of the new order. A few of the others, the "labor aristocracy," emerged as the designers and builders of the new machinery, while still others shifted from an old skill to a new skill and became machinists. Yet the earliest years of the Frankford Arsenal gave only the vaguest inklings of the changes just over the horizon.

Until after the middle of the century the enlisted workers remained in three familiar and easily identifiable categories. First came the master craftsmen—usually one master armorer, one or two other armorers, a master carriagemaker, and another carriagemaker or two. In addition to these highly skilled craftsmen, records occasionally identified a blacksmith or a carpenter. By 1839, when jobs had started to become more specialized, the Ordnance Department began to divide these same craftsmen into first and second "grades." The ancient term "artificer" applied to the next level of enlisted "soldiers of ordnance." These were also skilled craftsmen; indeed, the records indicate that these men were recruited specifically because they possessed a skill valuable to the Ordnance Department. Arsenal records identify a large variety of skills among the artificers. They include tinsmiths, beltmakers, wireworkers, harnessmakers, painters, and saddlers, among others. Frequently, artificers are listed as "sergeants of ordnance" who were not only skilled craftsmen but also more reliable men assigned by rank to supervise regular soldier/workers.

Both numbers and specific examples confirm this division of skills. The ratio of artificers to ordinary soldiers was almost always one to two, one to three, or at most one to four. In 1839 when the artillery at Fort Mifflin (located just south of the city of Philadelphia where the Schuylkill River flows into the Delaware) required extensive repair, the commander, Capt. George Ramsay, sent "artificer" William Robinson to supervise the work. During the Civil War, when it became impossible for either the commander or other commissioned officers to visit the many powder mills to inspect quantities of powder being readied for shipment, these artificers and sergeants of ordnance often acted as "sub-inspectors."

Finally, common soldiers identified as "laborers" held the lowest position in this hierarchy of workers. Often these laborers performed the menial jobs on the post. They made thousands of brushes and picks or sets of infantry equipment. They also labored out-of-doors maintaining the grounds and buildings, and they even worked as farmhands in the fields where hay was grown to feed the horses and other livestock on the post. When the Ordnance Department undertook major renovation or construction projects the commander often detailed these laborers to work on them. But these workers were just as likely to perform skilled as nonskilled tasks. As munitions technology evolved and experimental laboratories appeared, laborers served as the technicians in these shops. On November 5, 1861, the commander wrote:

> It is my painful duty to report the occurrence of a very serious and fatal accident at this arsenal about 2 o'clock P.M. today — an explosion took place in the building known as the mixing house.

The mixing house was one of these early laboratories. In it various chemical elements were mixed into a wet solution, then dried and remoistened slightly to be packed as gunpowder into cartridges. This operation required both skill and experience in handling dangerous chemicals. Two laborers, Patrick Cooney and Joseph Neal, died in this explosion.[21]

In addition to this solid phalanx of enlisted soldiers, various peripheral and temporary personnel augmented the work force of the arsenal. Because they served as a vital link in the contract system, the most important of these were the U.S. inspectors of arms. These were civilians with detailed knowledge not only of the manufacture of weapons, munitions, and accoutrements, but also of the tricks whereby shifty contractors passed off inferior goods for quality ones. Although the Ordnance Department itself and not individual arsenals employed these inspectors, they frequently operated from one arsenal under the supervision of the commander of that post. They were highly paid, usually at the same rate as the master armorers from whose ranks they came. These men also commanded authority and respect.[22]

On some occasions expediency required the hiring of temporary civilian workers for some large construction project. If the arsenal received an unexpectedly large shipment of muskets that required extensive repair, the superintendent (the post commander was sometimes addressed by this title) might "borrow" some gunsmiths or armorers from a local contract armsmaker such as George Tyron or John Rogers.[23]

For most of the first three decades a small and relatively stable work force performed the many tasks demanded of the Frankford Arsenal and the Ordnance Department. Some conditions such as hours appear to have been straightforward. A letter from a later period states, "The working hours at all Arsenals is regulated by Ordnance Regulations, approved by the Sect. of War. Paragraph 33 says, 'the working hours for hired men at the Ordnance Establishments shall be arranged as to average ten hours a day throughout the year, working by daylight only.'" Although these regulations referred specifically to "hired men," they might as readily have applied to enlisted workers. Thus, for a six-day work week (Monday through Saturday) the workers at the Frankford Arsenal kept hours comparable to other American workers in similar industries. Arsenal commanders remained aware of this parity and frequently inquired about conditions and wages in other local industries.

Soldiers of ordnance received a monthly salary regulated by the Ordnance Department and based on military rank. Yet here ambiguity appeared, for arsenal records indicate different rates of pay for different skills. In the second quarter of 1833 the master armorer received fifty dollars per month for a twenty-six-day month, averaging around $1.92 per day. In 1846 Lt. Andrew Dearborn sent to the Ordnance Department records of wages calculated over three years. An unspecified number of carpenters who had labored for 1,234.25 days received $1,718.89, which averaged out to about $1.40 per day. By comparison, less-skilled laborers had toiled for 804.25 days and received $733.94, or $.91 per day.[24]

We are left, then, with a less than conclusive indication of the wages of the arsenal work force in its early years, and other factors only complicate an already unclear picture. Soldiers of ordnance, like soldiers of the line, received in addition to their monthly wages certain other emoluments. They received through the Department of the Commissary General monthly rations of food, candles, and other items. They were also quartered in barracks built on the arsenal grounds. Through the quartermaster general's office they received fuel (the cost of which was deducted from their pay) and a yearly clothing allowance, which often went unused and seems to have been a soldier's nest egg. Records frequently indicate that at the end of an enlistment discharged soldiers received cash payments for unused clothing allotments. Finally, last on our list but first in the hearts of many soldiers, ancient prerogative entitled ordnance soldiers to a monthly whiskey ration. An 1828 circular letter from the Ordnance Department set the whiskey allowance at 31¼ cents per month.[25]

If wages and fringe benefits distinguished ordnance soldier/workers from other early nineteenth-century workers, one other aspect of the employer-employee relationship did not. The coming of the new industrial order, which physically and psychically separated employers and employees, created for the new capitalist class a problem novel to them, but ancient to the army: controlling a large number of men in subordinate positions. Since time immemorial military minds had struggled with and devised various stratagems to solve this managerial dilemma. These procedures—based on a system of rewards and punishments, used originally by military officers commanding enlisted soldiers, and then transferred to hired civilian workers at armories and arsenals—provided later industrialists with ready-made ways to instill order among large numbers of workers inexperienced in factory production. Ostensibly, military commanders possessed means (such as courts-martial, fines, and imprisonment) for controlling enlisted soldiers unavailable to ordinary employers. These heavy-handed punishments might or might not have effectively controlled ordinary soldiers, but they did not prove successful when applied to proud craftsmen. Instead the more competent commanders at the Frankford Arsenal, trained both as officers and engineers, instilled loyalty in their most competent and efficient workers by actively involving them in both the decision-making process and the technological innovations, as well as by paying attention to more basic concerns such as wages and working conditions.

Yet this more comfortable and efficient relationship between military employers and civilian employees evolved only slowly. From the beginning the Ordnance Department and the commanders of the arsenal clearly preferred enlisted soldiers to hired civilian workers, even though this system proved more costly because soldiers idled by seasonal interruptions of work still had to be paid, housed, and fed whereas hired workers could have been dismissed.[26]

Even with this more easily controlled detachment of enlisted soldier/workers, arsenal commanders experienced problems shared by other employers then and now. Lieutenant Dearborn agreed to the transfer of laborer Michael Kirwin, "provided he is not married and will take a vow of perpetual virginity." The commander of the arsenal at Fortress Monroe, Virginia, wrote to the commander of the Frankford Arsenal, "Artificer Avery got your ten dollars and ran away... he got $20 from me." And Lt. Col. William Worth requested permission to discharge John Morfit for "idleness and insubordination." However, intemperance and drunkenness appeared as by far the most common and recurring complaints. Lt. William Mellon notified the ordnance office that he intended to discharge William Reeds, "a confirmed drunkard whose family has been smuggling liquor onto the post"; and Lieutenant Thomas had earlier discharged an unnamed laborer as "a useless drunkard who had venerial [sic] disease."[27]

THE ARSENAL AND THE COMMUNITY

From the beginning the Frankford Arsenal served as the focal point for a complex set of relationships. Sometimes these many-layered connections appeared straightforward and obvious, but ambiguity and contradiction often marked the post's relationship with the local community. While some inhabitants saw the arsenal as a well-paying customer, others regarded it as an unwelcome military presence.

In February 1821 the commander, Lt. Martin Thomas, received a bill for $77.50 from the tax assessor of Oxford Township. After consulting both the U.S. attorney and the chief of ordnance, Lieutenant Thomas paid the taxes. Arsenal commanders continued paying until 1841, when the state finally ceded jurisdiction of the post to the United States government. Why the arsenal continued to pay local taxes after the Supreme Court declared such taxation unconstitutional in 1819 in *McCullough v. Maryland* and how termination of those taxes related to cession of jurisdiction by the state remain unclear.[28]

Other issues further aggravated the situation. In 1823 the township began improving the public roads that ran along two sides of the arsenal's property. Unfortunately, these "improvements" caused water to drain onto the grounds of the post in inclement weather. The commander protested but to no avail. In exasperation, he again contacted the U.S. attorney about commencing a suit against the supervisors of Oxford Township. The problem went unresolved until the Civil War.

Originally, a wooden fence secured the arsenal from intrusion. An 1832 woodcut shows a rural scene, several buildings surrounded by a split rail fence. Already, however, the neighborhood had begun to change. Colonel Worth wrote to the ordnance office,

> We are surrounded by Factories in which people of almost all nations and descriptions are engaged at work, these are more or less troublesome everyday.

In order to protect the arsenal the commander proposed building a stone wall with iron fencing. Once again, the neighbors objected and attempted to prevent construction of any such structure.[29]

In 1830 the Frankford Arsenal existed as one of the depots of the Ordnance Department of the United States Army. The arsenal and its small detachment of enlisted soldiers of ordnance performed a myriad of tasks, as the Ordnance Department attempted (against the prevailing antimilitarism) to keep pace with the rapid changes in European military technology. Although we know something about the business of the arsenal in 1830, the picture remains sketchy and incomplete. The paucity of records prevents surety. We cannot tell which activity—storage, refurbishing, or manufacture—dominated this early period, nor can we assess the continuousness of operations. Since we know that the Delaware River froze during some of the colder winters, we can guess that workers at the Frankford Arsenal labored at a seasonally determined rhythm.

Except for a few names, we know almost nothing about the lives of the people who labored there; they stand mute against us. Their lives, their hopes and aspirations, and their relationships and kinship networks lay beyond our grasp. This is what Peter Laslett refers to as "the world we have lost." Soldiers of ordnance stationed at the Frankford Arsenal appear in neither census records nor local tax records. They left no private records, no journals, few letters, and no diaries. Only in some rare instances, such as the case of Rebecca Deal Murray, do we see glimpses of this lost world.

On January 2, 1860, in a terse letter, arsenal commander Maj. Peter V. Hagner mentioned the death of the hospital matron Mrs. Rebecca Murray. Rebecca was the widow of Sgt. John Murray, who had served in the army since the War of 1812 and been reenlisted by several arsenal commanders. In 1851 records identified him as an armorer, a position of skill and trust, but he disappears from the rolls shortly thereafter. Rebecca was a member of the locally prominent Deal family. Joseph Deal, perhaps her brother, negotiated several construction contracts with various arsenal commanders. An 1822 list identifies Rebecca Deal as being admitted to membership in the Frankford Presbyterian church, the same church to which other members of the Deal family belonged. Here the records end. We are left to wonder about the courtship and marriage of this soldier of ordnance and this daughter of a prominent local family of wealth and status. Perhaps their marriage and life together could illuminate the relationship between the Frankford Arsenal and the community outside its walls—but both the important and the ordinary are lost to us, a part of history irretrievably gone.[30]

3

The Beginning of Change, 1831–1848

For a decade and a half the Frankford Arsenal existed as an intrusive presence in a long-settled community and a small and relatively obscure military outpost, part of a military establishment that was itself of little consequence. But the winds of change had begun to stir, and within twenty-five years they transformed the arsenal from an insignificant early industrial workplace into a facility on the cutting edge of changing industrial technology and a bastion of the increasingly potent American military establishment.

This transformation stands as a clear example of the complexity of historical change. Events in Europe, in America, and just outside the arsenal gates combined to influence this change. Alterations in American business, government, and society mingled with the needs of a rapidly expanding nation, an evolving federal bureaucracy, and the quickening pace of American inventiveness to affect the production of firearms and ammunition. New production methods irrevocably changed the Frankford Arsenal from a small collection of handicraft shops and warehouses into a giant industrial complex.

The European wars of the early nineteenth century produced growing dissatisfaction with inefficient weaponry. Both the European and American military communities began increased experimentation in the methods of obliterating life. French chemists, in particular, carried out complicated experiments in formulating better gunpowder. In both Europe and America innovators labored to improve the mechanisms for firing both large ordnance pieces and small arms. In the United States, the Ordnance Department played a major role in the creation of the American system of industrial production. In its armories and arsenals as well as through the contract system, the department promoted mechanization of manufacture, uniformity in production, and ultimately interchangeability of parts. Two key factors affected these changes. Increased bureaucratization and rigorous tests devised by the new West Point graduates, the soldier-technologists, enabled the Ordnance Department to hold both army personnel and civilian contractors to more exacting standards. Using these tools, the department initiated the industrial system of manufacture in America well in advance of private industry.[1]

The election of 1828 put Andrew Jackson, a dedicated militarist and bellicose imperialist, into the White House. Recent historical literature has challenged long-held notions about Jackson, the age of Jacksonian democracy, and his role in the changes that remade America in the 1830s and 1840s. Whatever Jackson's role in these wholesale changes may have been, his presidency forcibly affected the American military establishment. Jackson's belligerent foreign policy, the near conflict with France over spoilation claims in 1835, the dispute with Britain over the boundary of Maine, the attempt to purchase Texas from Mexico, and the forced removal of native Americans, along with the presence of several capable secretaries of war, promoted military growth and expansion.[2]

ORDNANCE DEPARTMENT BUREAUCRATIZATION

An expansion of the federal bureaucracy, evidenced in the paperwork churned out by the office of the arsenal's commander, paralleled this military growth. From the beginning the Ordnance Department insisted that arsenal commanders keep proper records. Among the earliest communications from the department in the records of the Frankford Arsenal is an 1821 memorandum that identifies the monthly reports that were to be submitted:

1—muster roll of hired men permanently employed including nature of employment and wages
2—same for enlisted men including a clear but succinct view of the Business going on at the Post
3—Monthly return of work performed by hired men . . . comparing the Wages of the Workmen with the product of their labor.[3]

The Ordnance Department also required quarterly and annual reports of a similar nature, except that the annual report also necessitated an inventory of all goods, buildings, and persons. Additionally, the commander had to estimate, by the end of one fiscal year, his expected expenses for the following year; by June 30, 1830, he was expected to project expenditures from July 1, 1831, to June 30, 1832. Other federal agencies besides the Ordnance Department also demanded endless reports and other submissions of paperwork. The Treasury Department audited post returns monthly, and both the quartermaster general and the commissary general of subsistence ordered similar accountability. By 1834 the Ordnance Department alone listed sixteen different forms to be submitted, including "Form 56 for Annual Reports." Long before the Civil War the ultimate device of bureaucratic domination, a form to complete to order other forms that then had to be filled out and submitted, had made its appearance.

Thanks to Maj. Sylvanus Thayer's reforms at West Point, the United States Army had come to understand the interrelated concepts of accounting, accountability, and management. At West Point, Thayer had taken unruly cadets and made them accountable for both their behavior (through the demerit system) and their academic work (through a system of daily recitation, grading, and monthly ranking). The officers who emerged from this rigorous regime then applied the same techniques to the management of both soldiers of ordnance and civilian workers, first at the Springfield Armory and later at other installations. Very early in the nineteenth century, the Ordnance Department had grasped and begun to implement one of the essential keys to successful industrialization.[4]

THE FIRST CIVILIAN EMPLOYEE

Since the commander of this small post often acted as both quartermaster and commissary of subsistence along with his numerous other responsibilities, paperwork often threatened to deluge him. To relieve the overworked commander the Ordnance Department eventually permitted him to hire a civilian clerk, the first permanent nonenlisted employee at the Frankford Arsenal. George Willard, an artificer and sergeant of ordnance, was discharged in July 1833 and hired as the clerk. He was paid $1.75 per day, making his salary equal to that of the master armorer. Willard held the position until 1838, when he retired and was replaced by William Pigott, another discharged soldier of ordnance who remained until 1862 and became one of the innovators of the new industrial order.

The importance of these civilian clerks greatly exceeded the respite that they gave to the beleaguered commander. As the arsenal evolved from a depot manned by handicraft workers to an industrial complex the role of the clerk evolved in tandem, and to some degree the increasing complexity of industrial activity can be gauged in the changing role of this clerk. First, the continually growing mountain of paperwork transformed this single clerk into a head clerk and office manager who by the end of the Civil War supervised twelve other clerical employees. Second, the clerk gradually assumed other and greater roles of both a managerial and a supervisory nature.

Most of this evolution occurred during the extended tenure of the second clerk, William Pigott, and will be discussed in subsequent chapters, but even the relatively brief stewardship of George Willard offered sufficient evidence of incipient change. From 1833 to 1838 Willard kept the superintendent's letter book in a (thankfully) neat and literate hand that gives some evidence of formal training, if not actual schooling—a fact that distinguishes him from other ordnance soldiers (the letters-received file of the arsenal contains enough letters written to the commander by ordnance soldiers to make a comparison). From this start Willard assumed more important duties. In the absence of the commander, which became more frequent as the tempo of arsenal business increased, Willard routinely made managerial decisions. He sent orders to contractors and compiled fiscal estimates of future expenditures for the Ordnance Department. When the department decided to remove its stores from the Schuylkill Arsenal and several other depots, the arsenal commander, Col. William J. Worth, assigned Willard to supervise a detachment of soldiers effecting the removal. When, in the midst of tension between the Ordnance Department and the state of Pennsylvania, it became necessary to retrieve arms and ammunition stored in the state arsenal in Harrisburg, the commander sent his trusted emissary Willard to supervise the removal, inventory the matériel, and handle the delicate negotiations with state officials. By 1838, the end of Willard's term as clerk, this position had clearly begun to evolve into one of more complexity and managerial responsibility.[5]

THE REORGANIZATION OF
THE ORDNANCE DEPARTMENT AND
THE ADVENT OF THE SOLDIER–TECHNOLOGIST

The military and bureaucratic growth of Jackson's presidency also resulted in changes in the Ordnance Department itself, which in turn directly affected the arsenal. In 1821, for economic reasons, the Ordnance Branch had been recombined with the Artillery Branch of the army. Congress reestablished it as an independent department in 1832, and three years later the secretary of war appointed Col. George Bomford as chief of ordnance. This renewed independence, coupled with the tenure of two dynamic and farsighted secretaries of war, Lewis Cass (1831–37) and Joel Poinsett (1837–41), infused the Ordnance Department with a sense of vigor that affected both the office in Washington and its far-flung outposts.[6]

Although this rejuvenation took many forms, the establishment of a permanent ordnance board stood as the hallmark of the new Ordnance Department. This board, generally composed of six officers, met twice a year to consider a variety of issues. Although it could raise questions of its own, problems posed by the secretary of war or the chief of ordnance usually served as the basis for its agenda. Since it reported directly to the secretary of war, the ordnance board avoided the internecine conflicts that often marred and retarded relationships among the various army bureaus.

From its inception to the Civil War the board considered and reported on a wide range of issues. For example, it debated whether bronze or cast iron was a superior metal for founding artillery pieces, and whether the Gribeauval or stock-trail carriage was more reliable for field artillery. It examined and initially rejected both Samuel Colt's revolver and Christian Sharps's breech-loading rifle because they failed to meet the three main requirements—"simplicity, serviceability, and dependability"—for weapons issued to soldiers in the field, but later it approved both weapons when the inventors, following its recommendations, made improvements to them.

During the 1840s and 1850s the officers of the ordnance board focused mainly, but not exclusively, on three specific technological problems: methods of testing and manufacturing gunpowder; improvements in small arms, especially replacing the old flintlock firing mechanism with the more reliable percussion system; and the manufacture of ammunition for both small arms and larger artillery pieces.

Even though the ordnance board and the Ordnance Department were often threatened by political machinations, a tight-pursed Congress, and an overwhelming volume of work, they succeeded because they adopted and held steadfastly to one main criterion in regard to new ideas: no invention, no matter how great the demand for it, would be adopted until it had been satisfactorily tested by verifiable and accurate scientific methods. This insistence on scientific accuracy marked the advent of the soldier-technologist and became the hallmark of the newly reorganized Ordnance Department.[7]

Thanks to Thomas Jefferson, the United States Military Academy at West Point was meant from its founding in 1802 to be a scientific institution. However, by 1817, when Maj. Sylvanus Thayer became superintendent, both the curriculum and the administration of the academy had fallen into disarray. Thayer, who had visited the famous French Academy, the Ecole Polytechnique, the year before he was appointed to West Point, immediately introduced reforms that transformed the school into a training ground for the emerging soldier-technologists. By the mid-1820s West Point began to produce graduates who had studied French (the language of European military technology), advanced mathematics, chemistry, physics, military drawing, and engineering—all courses designed to emphasize the scientific nature of warfare.[8]

In 1835, when the Ordnance Department reemerged from the shadow of the Artillery Department, a double handful of these young West Point graduates—all enthusiastic, trained scientists—stood ready to construct, test, and perfect the newest engines of war. Many of these able young officers went on to receive their baptism by fire in the war with Mexico and then capped their careers both as tacticians and field commanders on the bloody battlefields of the Civil War. However, years before either of these conflicts these young soldier-technologists changed both the American arms industry and their own arsenals and armories.

In October 1832 Lt. Col. William J. Worth arrived to command the Frankford Arsenal. Worth was not one of the soldier-technologists, for although he had graduated from West Point he had done so before Major Thayer's reforms; but he had served at the academy as the instructor of mathematics for many of these young officers. Changes took place immediately. Anticipating its coming independence, the Ordnance Department started systematically to remove its supplies from forts and other military posts controlled by other army branches. Acting on orders from the ordnance office, Worth sent parties that removed ordnance stores from the Schuylkill Arsenal, Fort Mifflin, and Fort Delaware and transported them to the arsenal. Consequently, as the Frankford Arsenal grew in importance and size the others diminished. The Schuylkill Arsenal, subject to flooding by the Schuylkill River, became less important. Forts Mifflin and

Delaware experienced brief reincarnations during the Civil War but disappeared shortly thereafter.[9]

The Ordnance Department also decided upon another change that fundamentally altered the nature of the work carried out at the Frankford Arsenal. From its creation the arsenal had built strong ties with E. I. duPont de Nemours Company of Wilmington, Delaware. Frequent large shipments of gunpowder, niter, and saltpeter were received at Frankford from duPont's works on Brandywine Creek. In 1835 a decision by the chief of ordnance, Col. George Bomford, and the secretary of war, Joel Poinsett, transformed those ties into unbreakable bonds. Those two administrators ordered that all gunpowder purchased by the Ordnance Department was to be proved at the Frankford Arsenal. From 1835 through the Civil War proving gunpowder consumed much of the effort of the arsenal commander. Often the commander traveled to Wilmington to inspect powder ready for shipment, but just as frequently the inspection took place at Frankford. Later other firms challenged duPont's dominance in gunpowder production for the American military, but the Wilmington firm remained supreme. Only duPont powder continually passed the increasingly rigorous tests imposed by the soldier-technologists. Furthermore, the duPonts seemed endowed with the gift of omniscience, for they possessed the knack of purchasing huge quantities of the imported ingredients for gunpowder when prices were low, but just before demand skyrocketed.[10]

A third significant change also marked Worth's relatively brief command of the post. In the estimated expenditures for fiscal year 1833, Worth's predecessor, Col. Joseph B. Walbach, had recommended a request for a congressional appropriation of $9,000 for "2 Brick buildings each 40 Feet long, 19 ft. wide." The Ordnance Department rejected the recommendation, and in October 1832 Colonel Worth replaced Colonel Walbach. Worth pursued Walbach's planned building program. On January 16, 1833, he petitioned Colonel Bomford for a "special expenditure" for two additional buildings, "one for a laboratory and the other for an Armory." Two weeks later Worth received a letter denying an appropriation of $2,800 for "the erection of two buildings." However, in March 1833 Colonel Worth's persistence was rewarded when Congress approved $5778.79 for "the buildings etc. at the Frankford Arsenal."

Having been successful, Worth pressed his luck on a grander scale. In the fall of 1833 he asked for another building. This was to be a three-story, 100-by-40-foot brick building costing $18,390. This time Worth succeeded on the first try. On May 19, 1834, Colonel Worth received notification from the chief of ordnance that Congress had appropriated $16,170 for construction of:

a . . . new arsenal at Frankford (120' × 40', 3 stories of brick and stone with slated roof), for repairing old arsenal & completing laboratory, removing and enlarging gun carriage shed; and for repairs to barracks, shops, fences, etc.[11]

One of the smaller buildings, the laboratory, indicated the future direction of both the Ordnance Department and the Frankford Arsenal. Its approval paid homage to the increasing impact of science and technology on the manufacture of munitions. Ammunition had to be made according to increasingly precise and scientifically verifiable specifications and by increasingly complex and rapidly developing technology. Having been successful in prodding the Ordnance Department into an extensive and expensive building program, Worth decided to press his advantage. He petitioned Colonel Bomford for an increase in the number of workers—by the expedient of temporary enlistment if necessary. Colonel Bomford curtly told him that under no circumstances could the number of workers be increased beyond the current limit of twenty-five.[12]

Unfortunately, Worth did not remain long enough to see the new arsenal finished. On January 12, 1835, he received another letter from Colonel Bomford ordering him to turn over command of the post to Capt. Alfred Mordecai.[13]

The arrival of Alfred Mordecai, perhaps the greatest of the soldier-technologists, marked the beginning of the technological revolution in the manufacture of arms and ammunition at the Frankford Arsenal. Mordecai, the grandson of a German Jewish immigrant and the son of a Hebraic scholar, had graduated from West Point in 1823 ranked first in his class. After graduation he had chosen the Artillery Branch, and he had spent considerable time supervising the construction of the fortifications at Fortress Monroe, Virginia. When the Ordnance Department was reorganized in 1832, Colonel Bomford chose Mordecai as one of the thirteen officers of ordnance and assigned him to the Washington Arsenal. Here Mordecai, already the rising star of ordnance technology who had received special notice from Secretary of War Lewis Cass, began serious experiments with arms and ammunition.[14]

Mordecai's appointment to the post of commander of the Frankford Arsenal gives evidence that the post had attained prominence in the eyes of the War Department. At Frankford, now the receiving site for all gunpowder, Mordecai pursued and brought to fruition improved methods for testing, or "proving," gunpowder.

Throughout its history the United States Army had clung to the use of the éprouvette system. In this antiquated and outdated method gunpowder was tested by loading a fixed mortar with a measured amount of gunpowder and a cannonball of specific weight. When the mortar was fired the distance that the ball traveled was measured. These distances were then equated with the strength and the quality of the powder. Inaccuracy and inconsistency plagued this method. When repeated under identical conditions, the distance covered by the same projectile could vary by several hundred yards.

First, Mordecai convinced the Ordnance Department to abandon this flawed method and to replace it with two devices, the ballistics pendulum and the gun pendulum, pioneered by French and British military engineers. The two devices were similar. In the former a standard ball was fired from a cannon loaded with a carefully measured quantity of powder at a target attached to a pendulum. The distance, calculated in degrees of arc that the pendulum moved, constituted the strength of the powder. The gun pendulum reversed this procedure. The gun was suspended, loaded, and fired; this time the recoil of the gun, again measured in degrees of arc, proved the quality of the powder.[15]

Having established technologically improved methods of testing gunpowder, Mordecai proceeded to conduct exhaustive tests on the physical and chemical properties of the powder itself. He studied the importance of the proportion of component ingredients, sulfur, saltpeter, and charcoal; the density of the powder; and the size and number of its grains. He persuaded the department that larger grains of powder better suited larger ordnance projectiles, while smaller grains suited smaller firearms and their cartridges. He demonstrated that the proportion of dust to uniform-sized grains was important, and he showed that windage — the variation between the diameter of the projectile and the inside diameter of the barrel — must be minimized. Then Mordecai devised a number of simple tests to provide ordnance inspectors with easier means of checking the quality of gunpowder. For example, the examination of the residue of a small amount of gunpowder burned on a copper plate gave evidence of the kinds and amounts of impurities that it contained.[16]

Like his predecessor, Alfred Mordecai commanded the Frankford Arsenal for only a brief time, three years. His biographer regarded his time there as a quiet hiatus in his career. From the point of view of Mordecai's rise within the ranks of the Ordnance Department this may be true, but in the history of the arsenal it is not. In addition to his experimentation, a number of other significant events marked Mordecai's tenure as commander. He completed the construction of the laboratory and new arsenal building begun by his predecessor — and with this completion the arsenal took on the appearance of a significant military establishment. The new arsenal stood at the eastern side of a square with most of the older buildings, including the commanding officer's quarters and the enlisted men's barracks, around the other sides. Mordecai took pains to beautify the parade ground with seventy-five dollars' worth of trees and shrubs, "conducive to the healthiness of the arsenal." He paid similar attention to the rest of the property by reflagging and graveling paths and walkways. After initiating an inquiry that included a search for the original deed, which revealed that the arsenal boundary included the whole bank of the creek to the low-water mark, Mordecai obtained approval from the Ordnance Department to have a local builder, Joseph Deal, erect a seawall along the creek bank to protect the low-lying arsenal grounds from seasonal flooding.[17]

Four years earlier the previous commander had noted the rapid change taking place in the vicinity and complained about the tide of immigrant aliens and surrounding factories. Obviously by the mid-1830s Frankford shared in the rapid industrialization transforming the Philadelphia region.[18] This incursion alarmed Mordecai as much as it had Colonel Worth. Captain Mordecai, however, finally convinced the Ordnance Department that the Frankford Arsenal ought to be both expanded and better secured. He received $2,000 from Colonel Bomford to purchase a narrow, three-acre piece of land, the Kennedy plot, that ran along the entire eastern boundary of the arsenal and $9,000 to build a protective wall. Mordecai submitted plans for a seven-foot-high wall of brick and iron. The department concurred with the design of the wall but substituted stone for brick as the basic building material.[19]

On August 7, 1838, Alfred Mordecai's brief but important command of the Frankford Arsenal ended when he was reassigned to the ordnance office in Washington, D.C. In 1840 Mordecai and three other officers made a trip to Europe, returning a year later with more innovative ideas from the advanced European military community. In 1855 Mordecai returned to Europe with George B. McClellan and Richard Delafield as American observers of the Crimean War. Mordecai's career had a bitter ending. To him as to other military officers the Civil War came as a stunning blow. A Southerner by birth with most of his family still living in the South, Mordecai found himself alternately cajoled and hounded by both sides. Finally, he resigned from the United States Army early in 1862. He remained in Philadelphia (where his wife's family resided) and lived out his life in relative obscurity.[20]

Alfred Mordecai had a lasting impact on the Frankford Arsenal. Under the direction of Secretary of War Joel Poinsett and a reinvigorated Ordnance Department, Mordecai changed the arsenal from an ordinary military depot to an installation preeminent in the testing of gunpowder and other munitions. It is tempting to enhance the significance of Mordecai's tenure as arsenal commander, especially the beautification and protective programs. Was the attention paid to the parade ground and paths an attempt to replicate the village green in a hostile, increasingly industrial setting? Is it an example of middle-class orderliness—is

Mordecai's brick and iron fence a military version of the Victorian picket fence? Are these reflections of a scientific, military mind or just of the mind of one politically and socially conservative army officer? All of these are interesting speculations based on only the most tenuous evidence, but Mordecai's career both as commander of the Frankford Arsenal and as one of the ordnance officers most responsible for the development of nineteenth-century American military technology stands on its own merit without speculative enhancement.

Although George Douglas Ramsay, Mordecai's successor as commander, also held a seat on the prestigious ordnance board and later served briefly as chief of ordnance, he was not a soldier-technologist. Instead he more closely resembled Col. William J. Worth, Mordecai's immediate predecessor. Ramsay graduated from West Point in 1820 and served with both the Corps of Light Artillery and the Corps of Topographical Engineers prior to his promotion to captain and appointment as the commander of the Frankford Arsenal. Like Colonel Worth, Captain Ramsay represented the best of the dedicated, energetic career army officers. However, George Ramsay's administration proved to be less technologically progressive, and once again events beyond the control of the commander or his superiors in Washington affected the arsenal.

Andrew Jackson, wreathed in the adulation of his grateful countrymen, had departed from Washington to retire to the peace of the Hermitage near the grave of his beloved Rachel—but he left behind a financial disaster. The Panic of 1837 ranks as one of the most serious depressions in American history, and it gravely imperiled the Frankford Arsenal. Faced with reduced congressional appropriations, the Ordnance Department cut its expenses by halting construction projects, curtailing unnecessary expenditures, and eliminating personnel. The regular army was already so small that reductions in its line units were generally impractical; the auxiliary branches—the Corps of Engineers, artillery, and ordnance—suffered the heaviest casualties in the war against red ink. The chief of ordnance notified Captain Ramsay that the number of enlisted soldiers of ordnance at the Frankford Arsenal was to be reduced to twelve. Ramsay immediately discharged two laborers and transferred an artificer to a line regiment; he expected additional discharges as enlistments expired.

The gradual economic recovery in the early 1840s brought little relief to either the War Department or the Frankford Arsenal. Under the administration of "His Accidency," John Tyler, administrative paralysis seeped into the War Department. In 1841 alone, three different men held the position of secretary of war. Even under the command of a vigorous officer such as George Ramsay, the Frankford Arsenal languished. The arsenal records that cover this period, 1841 to 1844, convey the impression of a post struggling against adversity.

UNREST IN AN INDUSTRIAL COMMUNITY

In 1844 one local event startled the Arsenal as other events of national scope never did. The Mexican hostilities came and went, marked mostly by an increasing flurry of activity. Similarly, arsenal records barely note the opening of the Civil War, the Battle of Gettysburg, and the assassination of Abraham Lincoln; only a brief ceremonial artillery barrage, followed by a hasty return to work, marked the end of the long and brutal conflict. But the Kensington Riots proved to be a different case.

In 1838 Captain Ramsay received the disturbing news that a mob had ransacked the state arsenal at Harrisburg. Concerned for the security of the Frankford Arsenal as the surrounding neighborhood rapidly industrialized, Ramsay urged the Ordnance Department to increase the detachment at the post in order to guard it adequately. The department had replied politely but in the negative. It told Ramsay that he could hire guards using the meager funds already appropriated, but that he could not increase enlistment.[21]

On May 8, 1844, Captain Ramsay hurriedly penned the following letter to Col. George H. Talcott, the acting chief of ordnance:

> Before this reaches you the public papers will have informed you of the riot raging in Kensington. The excitement is very great, and it is impossible to foresee the result. As I have not deemed it prudent to leave the Arsenal, I cannot speak from personal observation. . . . My neighbors are very much alarmed. . . . Fires are now raging over the ill-fated district. . . . Mr. H. Pratt has just returned from the scene of Action—he reports that St. Augustine Church and surrounding buildings . . . have been destroyed and that a large body of Irish are held in check in the wood on 2nd St. road by General [Robert] Patterson.

Two days later in another letter Ramsay expanded his concern:

> Indeed for some time past we have been annoyed by riotous parties from the City—and our heretofore quiet neighborhood has become of much disorder.[22]

Between these two missives Captain Ramsay had received an urgent request for weapons and ammunition from Morton McMichael, the sheriff of Philadelphia. Ramsay hesitated to fulfill the request, for he had been reprimanded for acceding to a similar plea after the earlier ransacking of the state arsenal. Ramsay sent the thousand stand of arms and fifty thousand rounds of ammunition only after receiving permission from the Ordnance Department. Later the same day Ramsay received another letter from McMichael, warning "that there is a rumor that rioters intend to attack the Arsenal and the Catholic Church at Frankford."[23]

Now sufficiently alarmed, the War Department ordered an infantry company from Fort Columbus, New York, to Philadelphia to protect the apparently threatened arsenal. Ironically this infantry company, Company "K," commanded by Capt. Charles F. Smith, turned out to be more disruptive than the rioters. During the first fifty years of the Frankford Arsenal, its records mention perhaps five courts-martial. Captain Smith came close to equaling that: in one year he disciplined, court-martialed, and discharged several soldiers, perhaps as many as seven or eight. The situation became so annoying that Colonel Bank, the commander of the Fifth Department of the Army, to whom Smith's requests were submitted, routinely began to deny them.[24]

THE MEXICAN WAR AND THE NEXT STEP

Finally, the coming of the Mexican war relieved the arsenal of this onerous burden. The War Department transferred Company "K" to Texas, and a flurry of activity spurred by the demands of war commenced at the Frankford Arsenal— but without Captain Ramsay. Knowing that combat experience offered the only hope of advancement in the army's entrenched officer corps, he requested and received a transfer to Texas. Ramsay left the arsenal under the command of a young second lieutenant, Andrew H. Dearborn, just as the post was about to be overwhelmed by the demands of war. Orders to let contracts and to rush shipments poured in. Quantities of gunpowder, ammunition, and various kinds of equipment were shipped both by sea and land and then up or down the Ohio and Mississippi rivers. Cargoes left the arsenal wharf and were transferred to oceangoing ships in the port of Philadelphia, eventually headed for Galveston, Texas, or Bernicia, California.[25]

Strangely, even though the demands of the Mexican war eventually thrust the Frankford Arsenal into the industrial age, at the same time the war deflected both the arsenal and the War Department from their more important future role. The eyes of the War Department turned westward. The Ordnance Department enlarged arsenals at St. Louis, Missouri, and Baton Rouge, Louisiana, and founded new ones at Fort Leavenworth, Kansas, and Bernicia, California. Orders for equipment for cavalry and horse artillery such as saddle blankets, harnesses, and carbine ammunition (a carbine is a shorter shoulder firearm used by cavalry soldiers) dominated the efforts of the arsenal. Thus, in spite of recent technological advances, the coming of the Civil War found the War Department and the Frankford Arsenal pointed in the wrong direction. The Civil War did not take place in the shadow of the Rockies, nor was it fought primarily by horse artillery and mounted warriors armed with carbines.

Even as late as 1842 the commander of the arsenal had responded to a query by emphatically stating, "This is not a manufacturing arsenal." But technological change—especially in the increasingly precise manufacture of small arms and ammunition—coupled with the demands of the Mexican war forced the Ordnance Department to march bravely into the industrial age.

Even during the bustle of the Mexican war Lieutenant Dearborn took time out to look around and note with alarm the changes occurring. He warned the Ordnance Department that "a manufactory on a very large scale [Nicholas Lennig's Tacony Chemical Works] is being erected on the Frankford Creek very nearly opposite the Arsenal." He urged the department to consider the purchase of the thirty-nine-acre Ashmead Farm that had just become available. Dearborn emphasized the importance of the farm's location between the arsenal and the Delaware River. The department balked at the asking price, $20,000, as exorbitant. Dearborn tried to assure his superiors that the price was reasonable and would be snapped up since "manufacturing sites . . . are in good demand here." He continued:

> During the present year there have been completed six extensive manufacturing establishments; cotton, Dye, Prints & chemical works etc. All of which have gone into operation within the last three months. All of these are within one and one half miles of this Arsenal; within the same distance there are now in operation some fifteen establishments of the same nature. . . .

Dearborn concluded prophetically, "These manufactories do not come singly but in crowds."[26]

However, the Ordnance Department, preoccupied with the Mexican war, dithered and procrastinated. Dearborn persisted. In his estimated expenses for the fiscal year ending in July 1848, he requested $75,000 for repairs, two new storehouses, and "purchase of the Ashmead property." He also proposed an armorer's and machine shop and a carriagemaker's and paint shop "with steam engine and fixtures" (this is the first suggestion of mechanization in the arsenal records). But again the department dragged its bureaucratic feet, and again Dearborn persisted. The following year he estimated precisely and emphatically:

For purchase of 38 acres 20,000
For 1 principal workshop to contain machinery for mechanical operation
for proposed Percussion cap establishment 6100
For 1 workshop for the preparation of Fulminate 2150
For 1 Drying House for fulminate 850
For steam engine Boilers & Heating apparatus . . . . 4500.

He concluded his estimate in prophetic terms. "I think a sufficient number of boys and girls can be hired in this vicinity to make 1,500,000 ball cartridges each month."[27]

Lieutenant Dearborn was not fortunate enough to witness the fruits of his persistence. With the conclusion of the Mexican war George Ramsay, now holding the rank of major, resumed his command of the Frankford Arsenal, and on January 9, 1850, he was told to expect a remittance of $20,000 for the purchase of Ashmead Farm. Although it would take four more years to construct the percussion factory and put it into operation, the Ordnance Department had reluctantly entered the industrial age.

By 1848 the Frankford Arsenal had weathered instability and uncertainty, and it stood on the edge of a period of continuous and dynamic growth. The reorganized and revitalized Ordnance Department had already taken the initial steps toward the creation of an industrial system for the production of arms and ammunition. It had instituted bureaucratic orderliness and introduced scientific methods of testing weapons and ammunition. The next stage in the history of the arsenal, from the end of the Mexican war to the beginning of the Civil War, would see these preparations brought to fruition.

4

The Coming of Mechanization, 1848–1860

The opening of the Mexican war found the Frankford Arsenal a small outpost, a part of a military establishment struggling against adversity; by the start of the Civil War the Frankford Arsenal was a very different place. In 1846 the regular army of the United States comprised just over eight thousand men, including both officers and enlisted soldiers. In the preceding two decades congressional penury and the prevailing spirit of egalitarianism had created an atmosphere of adversity for the army. Yet out of this adversity emerged a sleeker, more dedicated, and more professional army. From West Point, reformed by Maj. Sylvanus Thayer, had come a group of well-trained soldier-technologists to oversee the Ordnance Department's quest for uniformity and standardization.[1]

Since 1832 Col. George Bomford had commanded the Ordnance Department, for which these officers labored. Bomford himself symbolized the dichotomy inherent in an essentially conservative institution that had decided on change as a matter of policy. Seared by the debacle of the War of 1812 and impressed by the technological innovations adopted by the English and French military establishments, Bomford—a crusty, short-tempered man—and the Ordnance Department decided that change had become a necessity. But this change was not to be promiscuous and not for the sake of change itself. It would be directed toward uniformity, simplicity, and, in artillery and shoulder arms, interchangeability of parts. No change was to be adopted unless it met these criteria and was submitted to rigorous testing. This meant that the department had to begin by devising such tests.[2]

In the first half of the nineteenth century the Ordnance Department, rather than private industry, directed the evolution of the famous American system of manufacture. It pioneered both uniformity and interchangeability in manufacturing; its officers devised methods of testing. In consultation with private manufacturers these officers constructed the machinery that produced arms and ammunition. This technological evolution fostered division of labor among its workers, both civilian and enlisted.

Changing technology is not simply a matter of self-evolving machinery. Machines do not come into the workplace just because they represent a "better" or less expensive method of production. To be viewed holistically, changing technology must also be seen as a social process. At the Frankford Arsenal industrialization evolved and was fashioned out of the relationship between the soldier-technologists who commanded and the skilled soldier artisans, civilian machinists, and other civilian workers. Technological success at the arsenal came in an amalgam of men and machines.

Finally, as the arsenal developed into a large-scale industrial complex that employed a large number of civilian workers, the relationship between it and the local community changed dramatically. People in the vicinity no longer saw the arsenal as an unwanted intruder.

PRODUCTION OF SMALL-ARMS AMMUNITION

Before the Civil War, the Frankford Arsenal had taken the first of two steps in the direction of full-scale industrialization. This first step began in the shadow of an earlier crisis, the war with Mexico.

By the time of the Mexican hostilities a complex, interwoven set of changes forced the War Department to come to grips with the necessity of mechanized production of small-arms ammunition. The Frankford Arsenal already possessed a long and distinguished record for the hand production of such ammunition, and now, thanks to the efforts of its two most recent commanders, it stood prepared to expand its production beyond the manufacture of paper cartridge ammunition.

Since the early 1820s the arsenal had produced a variety of cartridges for various shoulder arms. Even though these cartridges were mainly the product of hand production, their manufacture had evolved into an elaborate, precise, multistep operation. Soon the arsenal began to produce a number of other items, for by the 1850s the ordnance board had tested and approved several important improvements in the firing mechanisms of both small arms and artillery ordnance.

Since before the revolution inventors and gunsmiths had tried, mostly unsuccessfully, to replace the flintlock and frizzen pan and the slow burning fuse as the firing mechanism on muskets and cannons. By the mid-1840s three improvements proved reliable. In 1807 a Scottish inventor, John Forsythe, had devised a percussion lock that replaced the awkward and unreliable flintlock with something that resembled a nipple or a small inverted cone. The most innovative part was the percussion cap. The cap, similar to a small thimble, was filled with an explosive, fulminate of mercury. To fire the musket the soldier placed a cap over the nipple and pulled the trigger. The hammer struck the cap, exploding the fulminate, which then fired the powder and ignited the gunpowder charge in the cartridge in the musket's barrel.[3]

A second alternative, Maynard's primer, vied with the percussion cap for the War Department's favor. Edward Maynard, a Washington, D.C., dentist and inventor, had long dabbled with firearms. Maynard's primer resembled the percussion lock in that it also eliminated both the flintlock and the frizzen, but it differed in other essentials. Ingeniously, Maynard had also devised an alternative to the percussion cap. He cut a very thin, narrow, and flexible strip of copper that was dimpled every quarter of an inch. Each dimple was filled with a small but precise quantity of explosive powder and sealed. Then the primer strip was carefully coiled into a roll. The Maynard primer resembles a roll of caps sold today for toy pistols. This roll was then inserted into the lock through a hinged panel in the lockplate. Pulling the trigger tripped a spring and a feed finger that advanced a charge into position under the hammer.[4]

Both the percussion cap and Maynard primer had imperfections. The percussion caps were so small that in darkness, cold, or other adverse weather conditions soldiers found them difficult to handle. The spring and finger mechanism required for Maynard primers proved to be so delicate that it occasionally jammed with the rough usage common to battle conditions. Cartridges still had to be loaded into the muzzle of rifles and muskets and then rammed into place.

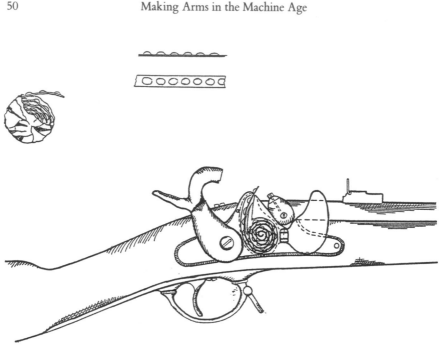

Fig. 10.   The Maynard primer and lock was an alternative percussion method. A coil of primers (shown above both in a packaged roll and as a partial strip) was inserted into the lock behind the hinged panel (shown open with the coil inserted). Once the musket was fired, a spring-activated feed finger pushed the next dimpled cap into position.

But even with these imperfections both the Maynard primer and the percussion lock were significant improvements. Shoulder arms fired more efficiently, more reliably, and with about half as much time between shots.[5]

The friction primer or tube, a third improvement, was for large ordnance pieces. Several American ordnance officers, particularly Maj. Theodore T. S. Laidley, cooperated on this improvement. The friction tube consisted of two tubes, a larger one filled with gunpowder and a smaller one filled with fulminate of mercury. The smaller tube was inserted perpendicularly into the larger one, and a rough copper wire was threaded through both. Yanking the wire by means of a lanyard exploded the fulminate, which ignited the powder to fire the cannon.

The Ordnance Department adopted these technological innovations slowly and with reluctance. Marked by a strong conservative streak, army officers seemed to fear rapidly firing armament—as if such weapons would engender a military promiscuity that would dilute control of battles as well as escalate the cost of warfare in both human and economic terms. But the outbreak of the Mexican war swept aside any such reluctance, creating a demand for both more efficient weaponry and vast quantities of munitions.[6]

From its creation as a separate branch of the army in 1812 until the end of the Mexican war in 1848, George Bomford commanded the Ordnance Department. Under Bomford's control two apparently contradictory qualities, steady growth and stubborn conservatism, characterized the bureau. For ten years prior to Bomford's death on March 25, 1848, another officer, Col. George Talcott, had supervised the routine administration of the office. Following Bomford's death, Secretary of War George Crawford appointed George Talcott as chief of ordnance, and two months later Talcott was promoted to the rank of brevet brigadier general.[7]

Significant change followed Talcott's appointment and promotion. He took steps to implement improvements that the ordnance board had advocated since its origin ten years earlier. These innovations varied from the very specific to the very general, from the procedure used to harness artillery pieces to teams of horses to more scientific and technologically advanced production methods for arms and ammunition. When implemented, these changes affected the whole Ordnance Department, individual officers, enlisted soldiers, and its numerous installations including the Frankford Arsenal.

Since 1847 various arsenal commanders had suggested to the ordnance office the expansion of the arsenal property and its production and research facilities. Now the Ordnance Department began to implement both suggestions. On January 9, 1850, Maj. George Ramsay, who had resumed command of the arsenal at the conclusion of the Mexican war, received the long-awaited remittance of $20,000 from the chief of ordnance to purchase Ashmead Farm, adjacent to the arsenal. With this purchase the Ordnance Department doubled the area of the post and cleared the way for significant expansion.

Within two months of purchasing the additional property, Major Ramsay negotiated another contract with a local builder, Joseph Deal, for a new wall to enclose the old farm within the arsenal's secure perimeter. Unlike some of his Frankford neighbors, Deal welcomed the arsenal's lucrative presence.[8] This contract signaled the beginning of the change in the relationship between the Frankford Arsenal and the surrounding community. Over the next twenty years this relationship evolved from one of hostility to one of interdependence.

By 1850 Frankford had long since left behind its rural past. Increasingly, commanders had become alarmed by transformations in the vicinity of the arsenal as the area rapidly industrialized. Factories sprang up seemingly overnight, and troublesome people filled the neighborhood. Arsenal records show the sense

FIG. 11. The stone wall built to keep out intruders in the 1850s, when the arsenal's rural surroundings evolved into an industrial community.

of alarm that these incursions evoked in some people. To justify the expense incurred in building a stone wall one thousand feet long and six feet high, Major Ramsay wrote:

> This neighborhood has become . . . for lawless acts, for many months scarcely a week has elapsed without the burning of houses and barns immediately about us by organized bands of desperadoe [*sic*].[9]

After completing the preliminary survey for the wall George Ramsay turned his attention to more important matters, the erection of workshops on the newly acquired land. In his estimate of expenses for the first quarter of 1851, Ramsay proposed "3800.00 [dollars] . . . for erection of two workshops each 26' x 65'. . . ." The Ordnance Department approved with unusual alacrity, for on May 26, 1851, Major Ramsay entered into another agreement with Joseph Deal—this time for the construction of two buildings, "each sixty-five feet by twenty-six feet, one story brick with slate roof."[10]

Politics had always played an important role in the evolution of the arsenal, but now national political intrigue intruded once again. In the summer of 1851 George Talcott's brief but stormy tenure as chief of ordnance came to a sudden end. Talcott had engaged in a running battle over political appointments in the ordnance service with Charles Conrad, Millard Fillmore's newly appointed secretary of war. In retaliation for Talcott's obstinancy, Conrad ordered an investigation of all procurement contracts and used the findings not only to dismiss Talcott from his position as chief of ordnance but also to drum him out of the army.[11]

Henry Knox Craig was born on March 7, 1791, in Pittsburgh, Pennsylvania, the son of Maj. Isaac Craig, a revolutionary war officer. Unlike some other contemporary ordnance officers, Craig (who had been in the army since the War of 1812) was not a graduate of West Point and had worked his way up through the ranks. On July 10, 1851, Col. Henry Craig, a hero of the Mexican war, replaced George Talcott as chief of ordnance. For the Frankford Arsenal, Craig's appointment proved fortunate. At a time when Congress was reducing the army's budget, Craig, who had commanded the arsenal briefly during the summer of 1845, succeeded in securing funds for continued research and development of weapons and munitions.[12]

Other changes followed quickly upon Craig's promotion. Colonel Craig transferred Major Ramsay to command of the St. Louis Arsenal, and on September 17, 1851, Maj. Peter V. Hagner replaced Ramsay as commander of the Frankford Arsenal.[13]

Hagner, a native of Washington, D.C., had already served at the arsenal as a junior officer in 1838, and just before his elevation to command he had been stationed at the Washington Arsenal and had become familiar with the new machinery used there in the manufacture of percussion caps.[14] Within three days of his appointment, Hagner wrote to the Ordnance Department concerning plans to build a "cap factory." In these letters he also inquired about the availability of some machinery in use at the Washington Arsenal and other arsenals. Following this inquiry, he submitted plans for a steam-powered factory to join the two workshops for which Major Ramsay had previously contracted. By February 1852 Hagner received permission to construct the percussion cap factory and several other buildings as well. Hagner designed the cap factory as a brick building (one hundred feet by thirty feet) to join perpendicularly the two workshops, giving the whole complex from an aerial view the appearance of a stubby letter "H."[15]

FIG. 12. Building 62, the percussion cap factory. Except for the slate roof, this building was altered extensively and no longer retains its original H-shape. The brick arches are all that remain of the many large windows that once lined the right wall.

Included among the other buildings were a "French" laboratory and a new warehouse to store niter. In the laboratory soldiers of ordnance processed gunpowder and conducted chemical experiments. The term "French" paid homage to that nation's role in early chemical experiments relating to explosives. This laboratory and three other frame buildings were constructed a short but safe distance from the cap factory. The storehouse for niter, on the other hand, was constructed as far away from all other buildings and as close to the river as possible. In December 1851 Major Hagner had forwarded to Colonel Craig a "special report" indicating that a "minute" inspection had discovered that more than two million pounds of niter—some in casks for twenty years—was stored in four warehouses situated around the heart of the compound. Since an explosion of this quantity of niter would both obliterate the post and significantly alter the topography of a large portion of the Delaware River valley, it seemed desirable to locate the new, large niter storehouse (one hundred sixty feet by fifty feet) at the extremity of the post, even though this entailed moving the dangerous substance over half a mile. Confidently, Hagner reported that he expected the buildings to be completed by August 1852.[16]

Having received permission for such large-scale construction, Hagner hastened to the next step, assembling the necessary machinery. On April 2, 1852, he reached an agreement with J. T. Sutton, a Philadelphia firm, to manufacture a ten-horsepower steam engine. In July the ordnance office granted him permission to visit Springfield Armory, the arsenals at Watertown, Massachusetts, and Watervliet, New York, and other private establishments "to examine percussion cap machines and others that may be advantageously transferred to the manufactory at Frankford Arsenal." Hagner compiled a shopping list that included a planing machine, a lathe (records do not identify this as a Blanchard lathe for irregular turning, but the arsenal soon commenced working with muskets in a fashion that would make this type of lathe desirable), and a large number of other machines from the Watertown Arsenal:

1 Pair of Cap Machines-the last made
1 Pair of Filling Machines-the last made
1 Machine for cutting copper
3 Machines for parts of friction tubes
1 Lathe with Chucks, Slide Rest + Tools for repairing
1 Small Drill Press.[17]

With buildings nearing completion and machines being gathered, the Frankford Arsenal still lacked one of the essential ingredients of an industrial complex, the factory worker. An industrialized work force evolved slowly at the arsenal and did not come to exist fully until late in the Civil War, but in 1852 Major Hagner began to take steps to assemble this work force.

THE BEGINNING OF A NEW WORK FORCE

Even before Major Hagner started assembling his new work force in 1852, several earlier workers had taken advantage of the growth and the evolution of the arsenal's business to advance themselves. However, since these changes came before mechanization, they can be cited only as precursorial. The case of George Willard and his evolution from artificer of ordnance to supervisory clerk has already been examined. Two others, William Adams and William Pigott, followed

George Willard's path into successful employment in the emerging bureaucracy of the new industrial system.[18]

William Adams first appeared in the arsenal records in June 1839 as an enlisted laborer, a harnessmaker and saddler, being sent to Camp Washington, New Jersey, on an errand to the New Jersey militia. In February 1840 the commander, Capt. George Ramsay, reported that a terrible accident had taken place in the blacksmith's shop. A musket, held in a vise while being repaired, accidentally fired and critically wounded the worker at the adjacent bench. The injured soldier was Adams. As Adams lay near death, Ramsay, in several letters, railed against both the Ordnance Department and the Quartermaster's Department about the lack of medical attention and mattresses in the post infirmary. By circumstances of which we are not apprised, Adams managed to recover. In the fall of 1842 Ramsay began to process the discharge of the recovering but crippled artificer only to discover that soldiers of ordnance were ineligible for pensions. Over the next year Ramsay worked ceaselessly, even petitioning Congress, to obtain a pension for Adams. Finally in 1843, with the permission of Colonel Talcott, Ramsay resorted to another strategy. He received permission to re-enlist Adams provided that he certified Adam's physical ability to the adjutant general's office. When Ramsay was transferred to Texas during the Mexican war he requested and received permission to have Adams, now a sergeant of ordnance, accompany him. While in Texas Adams devised a new cartridge box that was later patented and officially approved by the ordnance board. With the conclusion of the Mexican war Adams returned to the Frankford Arsenal and was discharged and then rehired, again by Ramsay, as a master armorer. In 1850 William Adams disappeared from the records of the arsenal; in 1863 William Adams held the important position of military storekeeper at the mammoth Washington Arsenal.

Adams's case demonstrates only that an individual could, under unusual circumstances, flourish within the Ordnance Department. It does not necessarily show an individual advancing within the maelstrom of industrialization. The case of William Pigott, on the other hand, does.[19]

While William Adams left the Frankford Arsenal before Peter Hagner assumed command, William Pigott remained, holding the position of overseer of shops and clerk. In 1852 Major Hagner added three other civilians, George Wright, George Esher, and Robert T. Perkins. These four (a fifth man, Robert Bolton, was added in 1857) comprised the hierarchy of the new industrial order.

William Pigott first appeared in the arsenal records in 1835 described as an artificer of ordnance, but by 1839 he had been discharged and rehired as a civilian clerk. The first hint of Pigott's competence came in that year. When the master armorer suddenly left, Pigott was assigned to supervise the repair of ten thousand

Springfield muskets. Pigott's talents transcended mere clerical ability. Within two years records identified him as clerk of the arsenal. In 1845 as Maj. Henry Craig was leaving command of the arsenal, he pleaded Pigott's case to the chief of ordnance. He asked and received permission to promote him to "Overseer of Shops and Clerks" with a salary of $2.25 per day, a dollar a day more than other skilled craftsmen. In arguing for Pigott's promotion Craig cited his "superior local knowledge" among other attributes. A letter dated August 5, 1848, indicates the position to which this former soldier of ordnance had risen. William Pigott had been sent to the office of Major Crossman, the quartermaster general stationed in Philadelphia, on arsenal business and had been treated badly. In the letter Major Ramsay, the arsenal commander, severely reprimanded Crossman for his treatment of Pigott, noting that Pigott was "entitled by character and position to respect and consideration."

By 1857 his salary had been increased to $2.75 per day, making him the second highest paid civilian employee. Four years later at the outbreak of the Civil War another commander, Lt. Thomas J. Treadwell, also strongly recommended that the chief of ordnance again increase Pigott's salary, writing, "His labors are very great, now often necessitating work at night and on Sunday."

Just when Pigott's position seemed most secure a crisis occurred. On October 7, 1861, an apparently unexpected letter arrived. Signed by Secretary of War Simon Cameron, it stated, "The Officer in command at the Frankford Arsenal will please appoint Lewis M. Troutman to the clerical position now occupied by Mr. William Pigott." The alacrity with which Lieutenant Treadwell responded to Pigott's danger stands as evidence of Pigott's character, the esteem in which he was held, and his importance to the arsenal. In a letter to the new chief of ordnance, General Ripley, and not to the secretary of war, Treadwell wrote:

> it is presumed that the appointment of Mr. T. was made under the impression that Mr. Pigott's health would not allow him to resume the duties of his office, Mr. Pigott's is now restored, and he can resume the duties of his post. . . .
>
> Mr. Pigott's long and faithful service at this Arsenal is well known. . . . I respectfully request that if the appointment of Mr. Troutman cannot be reconsidered that I may be authorized to retain Mr. Pigott as Superintendent of Stores, at his present rate of pay.

Within two days General Ripley acceded to his urgent request.

Somehow Pigott managed, with the help of friendly superiors in the ordnance service, to weather this crisis. When Edwin McMasters Stanton replaced

Simon Cameron and Gen. Henry Craig, who had commanded the arsenal while Pigott was chief clerk, replaced General Ripley, Troutman's appointment was rescinded and Pigott was restored to his position as chief clerk, a position that he continued to hold until he retired later in 1862. Pigott's career was long and distinguished, but his success cannot be attributed directly to the evolution of mechanization in manufacture of small-arms ammunition at the Frankford Arsenal. Rather, Pigott's successful career stands as evidence of the possibility of social and economic mobility, at least for some workers, in jobs created on the periphery of increasingly complex industrialization.[20]

George Wright had held the position of master armorer at the Washington Arsenal. While in this position Wright had devised the machine to manufacture percussion caps that bears his name. In fact, Wright's cap machine mainly improved on an earlier machine invented and used by French munition makers. The French machine funneled the tiny percussion caps into a chamber in the machine where they were filled with a minute explosive charge. Workers completed the laborious process of placing them in trays by hand. Wright's machine joined the two steps into one continuous mechanized production. The machine (see Figure 13), originally operated by a hand crank, was readily adapted to steampower. In 1852 Wright transferred to the Frankford Arsenal and, although he retained the ancient title of master armorer, became the master machinist and foreman of the cap factory. In addition to his mechanical skill, Major Hagner identified Wright as "a first class tin, copper, sheet iron, and brass worker. . . . " In his new capacity his salary increased from $1.75 to $2.00 per day. As an added incentive Hagner leased him a farmhouse on the arsenal grounds, approximately 200 feet from his workplace, for six dollars per month.[21]

George Esher's background remains unknown. It has not been ascertained whether he was a soldier of ordnance, a civilian employee at another arsenal, or a local worker. Perhaps he was recommended by the manufacturer of the steam engine. Major Hagner hired Esher at $1.45 per day to be the attendant of the new steam engine, but he also noted that Esher was "a good forger and filer and sufficiently handy at other work to assist where needed." Later Esher's salary was raised to $1.75 per day when it became obvious that steam-engine attendance was a round-the-clock job, especially in freezing weather, whether or not the machinery was in operation.[22]

The description of Wright's and Esher's skills furnishes us with an important clue in solving the riddle of the "deskilled" industrial worker described by some historians. Esher and Wright not only developed new skills related to the coming of mechanization, but also possessed old skills that had clearly retained value even as mechanization advanced.[23]

Robert Perkins, the most important of the three new civilian workers at the arsenal, completed the new industrial triumvirate. Perkins had held the position of master armorer at the Watertown Arsenal. Since the ordnance service was a small, relatively closed society, it is not surprising that Major Hagner knew of Perkins and his skills. In May 1852 Hagner asked permission from the ordnance office to contact Perkins regarding a transfer. Both the office and Perkins must have acquiesced, for later that month Hagner communicated directly with "Mr. Perkins" regarding supervision of the construction of the boiler and the installation of machinery in the new cap factory. As evidence of the high regard that the commander had for Perkins's ability, Hagner also consulted him concerning new lathes and planing machines not available from other machine makers. Within nine months Hagner began referring to the facility as "Mr. Perkins' new shop" and boasted about being able to make "very accurate screws" and "marking rules for drawing . . . more reliable."[24]

Perkins not only served as the superintendent of production and collaborated with Hagner on the design and construction of new machinery, but he also acted as Hagner's general factotum. When local manufacturers contracted to build machinery Hagner generally sent Perkins to inspect the product, obviously relying on his judgment. Later, one of these contractors wrote to Hagner accusing Perkins of trying to lure workers from his works. If Perkins had been guilty of aggressive industrial recruitment, there is no evidence that he was reprimanded for it. Perkins remained in Hagner's favor. In 1857, at Hagner's request, the ordnance office raised Perkins's salary to $3.50 per day, seventy-five cents more than William Pigott's. Until he retired in 1863, Perkins remained the highest-paid civilian employee at the Frankford Arsenal.[25]

Later a fourth "labor aristocrat," Robert Bolton (sometimes spelled Boulton), joined the threesome of Wright, Esher, and Perkins. In October 1857 Bolton, who in 1839 had invented and patented an earlier version of a cap machine, wrote to Major Hagner, with whom he was apparently acquainted, from a local address and stated, "I have received a situation which I have accepted temporarily, until you should want my services. I am very desirous to have a berth so that I can be at home with my family and one that would be steady." The next year Bolton's name appeared in the arsenal records as master armorer. When Robert Perkins left the arsenal in 1863, Bolton succeeded him as superintendent of industrial production.[26]

Here, then, were four men who adapted to and flourished with the coming of mechanization. These workers enhanced old skills that continued to hold their value and developed new ones as well. Clearly, the advent of industrialization did not automatically deskill all the workers emeshed in its grasp.

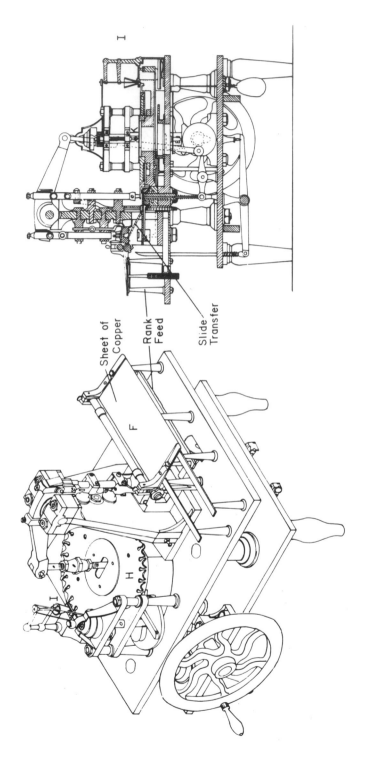

Sheet of Copper

Rank Feed

Slide Transfer

FIG. 13. George Wright's cap machine combined two previously discrete steps. At right, a worker inserted a sheet of copper (F). Turning the crank activated the double punch and dropped the unfilled cap into the rotating plate (H). The funnel at left (I) inserted the charge of powder to finish the process. A flywheel and belt later connected the machine to a steam-powered shaft.

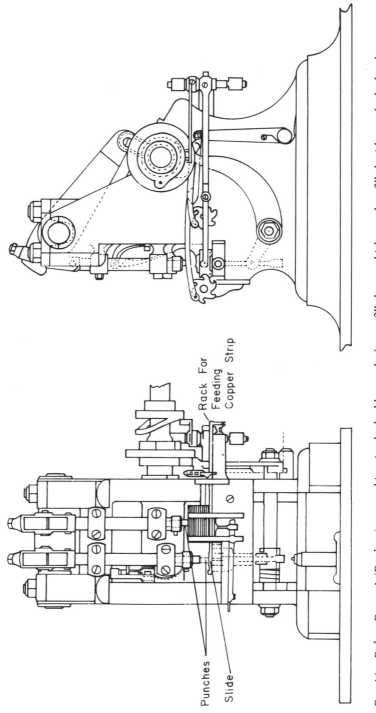

Punches

Slide

Rack For
Feeding
Copper Strip

Fig. 14. Robert Bouton's (Boulton) cap machine simply double-punched an unfilled cap, which was then filled with powder by hand.

If these four men stood at the top of an increasingly elongated and divided work force, what about the bottom? Both the records of the arsenal and the ordnance manuals refer to the employment of boys in munitions manufacture. Most writers, historians and nonhistorians alike, have generally regarded child labor as one of the greatest evils of industrialization. The records of the Frankford Arsenal suggest that another interpretation is possible.

In 1848 an auditor in the Treasury Department had disallowed payments that the commander, Lieutenant Dearborn, had made to "minors." In defending the payments, Dearborn listed the names of the boys and the amounts to which they were entitled. The names of the nine boys suggest that they were related to enlisted men stationed at the post. Early in 1848, two boys, Hiriam Dungan and William Owens, had received $9.75 and $7.22 respectively. Charles Dungan had enlisted as a laborer in 1844, and unless he was discharged early (of which there is no record) he should have been on the post in 1847. John Owens had enlisted as a laborer in 1831 and had been promoted to armorer in 1839. Even though his arm had been crushed in an accident in 1846, Owens, then forty years old, was recommended by the commander for the position of master armorer and "machinist." In 1851 Owens reenlisted for a fifth time. This required special permission due to his infirmity. In January 1852, in recommending Owens's reappointment as master armorer, Major Hagner cited his qualifications: "I believe the department has few men worthy of such promotion. He is now over 40 years of age, has a large family. . . . " Owens continued in the ordnance service until his death on February 9, 1865. His wife, Ellen, survived him.[27]

Like other arsenal commanders, Major Hagner both defended the practice of hiring boys and defended the boys as well:

> The boys employed in making Friction tubes become every year more experienced and their labor more valuable. As they grow older too, their labor should yield them more in order to secure their services and reward their attention to duty.[28]

The employment of boys at the arsenal suggests something other than industrial exploitation. It offers evidence of enduring vestiges of familial strategy and an artisanal past. Through employment of a son at the arsenal, a large family such as that of John Owens could have augmented the family income, or a father could have initiated the son into a skilled occupation that would guarantee future productive employment. Whatever the case, the experience of the men and boys described here strongly refutes the notion that the rapidly industrializing work force constituted a universally exploited mass. It does, however, support the picture of a work force stretched longer from top to bottom and marked by greater specialization and division of labor—and this stretching occurred more rapidly as the pace of industrialization quickened.

THE NEW PRODUCT

By February 1853 the new shops had opened and the production of percussion caps had begun. Even though machines produced the percussion caps, much of the setup and finishing work still required hand labor. First, skilled workers prepared the powder with which the caps would be filled. This procedure was the same as that of powder for cartridges. This was dangerous work that required both skill and at least rudimentary knowledge of chemistry. Readying the sheets of copper for production was the second preparatory process.

> The copper is cleaned by immersion in a pickle made of 1 part (by measure) of sulphuric acid and 40 parts water; it is scoured with fine sand and a hand-brush, and washed clean in running water,—after which it is well dried in clean sawdust and rubbed over with a cloth slightly oiled: it is then ready for the machine.[29]

Note that this process required a particular type of skill and judgment. The worker had to judge when the copper was clean, when it was "well dried," and what constituted "slightly oiled." Fine skills persisted in the middle of mechanized production.

Now the copper was "ready for the machine," which performed several operations:

> The sheet of copper (14 in. by 48 in.) is adjusted on the table of the machine. The hopper is filled with the percussion powder and the machine put in motion. The star or blank is cut by a punch and transferred to a die, where it is formed into a cap by a second punch. The cap is caught in the notches of the revolving horizontal plate, and carried first under the hopper containing the percussion powder, where it receives its charge of ½ grain, and then under a punch, which presses the charge firmly into the cap, and lastly to the drop-hole, where it falls into the receiving-draw.[30]

The last two steps in the process again necessitated hand work. A worker inserted a drop of shellac-varnish into each filled cap. This both sealed and waterproofed it. Later, this was also done mechanically by a drip-wick under which the tray of filled caps passed at the end of the machine. Finally, a boy counted and packed the finished caps, first into bags of ten thousand and then ten bags to a wooden box, "length 28.75 inches; width, 12 inches; depth, 8.5 inches."[31]

The ordnance manual did not specify the total number of master, workers, and boys involved in percussion cap manufacture, but the process must have required nine to twelve men and boys. On the other hand the manual did estimate time and quantities. "The average work day of ten hours, including all necessary stoppages is 31,000 caps for each machine." By the fall of 1853 the workers at the Frankford Arsenal had exceeded these estimates. In November 1853 Hagner reported producing "1,000,000 percussion caps, 25,000 friction tubes, and 100,000 Maynard primers."[32]

THE RIFLING MACHINE
AND THE REMINGTON CONTRACT

Much of the literature on the American system of manufacture and other technological advances made in early American industry credits the creation of this system to private manufacturers such as Eli Whitney, Samuel Colt, and Eliphalet Remington. This study has already drawn attention to the role of the Ordnance Department in the rapid industrialization and technological mechanization of small arms and ammunition production. The episode of the rifling machine and the Remington contract presents an even sharper contrast between the measured but steady technological advancement of the Ordnance Department and the groping reluctance of private industry.

By 1853 the percussion cap factory was turning out more than one million percussion caps per month, greatly exceeding the demands of the peacetime army. Feeling that the newly operational machinery was underutilized, Major Hagner began searching for work. He fell upon the idea of altering the older flintlock muskets in the army's possession over to percussion, and rifling their barrels to accommodate the minié bullet. On November 9, 1853, Major Hagner received permission from the Ordnance Department to alter two thousand old muskets. This necessitated two separate changes: first, replacing the old flintlocks with percussion locks, and second, rifling the barrels. Hagner started to search both for machines and for manufacturers capable of the dual undertaking.[33]

Over the next year Major Hagner visited the armories at Springfield and Harpers Ferry and communicated with several manufacturers, including the Ames Machine Company of Chicopee, Massachusetts, and the Remington Arms Company of Utica, New York. In November 1854 he negotiated a contract with the Remington Arms Company for two thousand percussion locks, complete with necessary parts. As part of the contract, Robert Perkins, Hagner's trusted emissary and master machinist, spent more than a month in Utica supervising the initial operation. The contract also stipulated that a "sub-inspector," John Taylor, from the Springfield Arsenal would remain at the Remington works and inspect the locks as Remington finished them. The locks were to be produced in lots of two hundred and shipped to the Frankford Arsenal. Taylor was equipped with an intricate set of gauges to inspect the locks; his salary was to be paid by the Ordnance Department.[34]

For the rifling of musket barrels, Major Hagner adopted a different strategy. One can infer that he had been dissatisfied with rifling machines that he had inspected. Such machines cut the spiral grooves in the interior of the barrel by inserting rods tipped with cutting heads. Cutting the grooves to the precise depth required several successive, time-consuming operations. Hagner, a techno- logical innovator of the first order, devised a machine that reduced the several operations to one. He invented a rod with multiple cutting heads that cut the grooves in one pass. He further improved the machine by having it cut three barrels at a time rather than one, and two years later he made it self-acting.

However, when he attempted to have the machine built he ran into a stone wall of industrial conservatism. No manufacturer, not even the vaunted Ames Company, was willing to undertake the extensive retooling required to build the machine to the specifications that Hagner demanded. Hagner searched for more than a year before he located a manufacturer willing to attempt the project. Ironically, the manufacturer that Hagner eventually located, the Bridesburg Machine Manufacturing Company, was a neighbor of the arsenal.[35]

By March 1856 the rifling machine was in operation and Hagner crowed at its success. "My rifling machine is becoming daily more perfect in its works and more expeditious so that I can promise for it 30 barrels per day of ten hours. . . . " Within another year the efforts of the workers increased the output to "39 or 42 barrels in the working day 8½ hours."[36]

Yet even with the success of the machine, manufacturers declined to produce others. The Bridesburg Machine Manufacturing Company claimed that it had lost money at the enormously high price of $800 on the first machine. Another prestigious Philadelphia firm, Nathan Sellers and Sons Company, also declined, saying "they are incapable as yet of producing heavy machinery to the degree of accuracy required." Even though we are unable to assess accurately the reluctance of several manufacturers to build Hagner's rifling machine, it is tempting to conclude that as late as 1857 the technological progress of the Ordnance Department exceeded that of private industry.[37]

By contrast with the success of Hagner's rifling machine, the failure of the Remington contract looms large. In August 1855 the arsenal received a shipment of two hundred locks from the Remington works. Immediately, complaints began to surface about the quality of the work, especially the breech pins. Hagner complained, "the parts for inspection [are] coming in very slowly and irregularly." In February Hagner began to prepare an abstract to the Ordnance Department about the poor quality and irregularity of Remington's production. "No delay at all significant can be ascribed to anyone but the Contractors . . . solely due to their own shops." Hagner, an apparently patient man, even lost his temper and complained bitterly to Eliphalet Remington, "I have been obliged to dismiss my filers for 3 days past" and "shall have only enough [breech pins] for ten days and will I fear have to suspend the stockers work." Both the irregularities and the complaints continued for more than a year. Finally, in April 1857, Major Hagner received a letter from the Ordnance Department to the effect that Remington wished to abrogate the contract, having completed less than half the work. By March 1858 Remington ceased manufacture of percussion locks.[38]

The Ordnance Department released the Remington Company from its contract without recriminations. Both parties seem to have realized that the technological precision demanded by the department exceeded the industrial capacity of the company. Within several years the Remington Arms Company successfully fulfilled other War Department contracts, and by the end of the nineteenth century it had become a major producer of arms for the U.S. government.

The episode of the rifling machine and the Remington contract demonstrated what Merrit Roe Smith has suggested—that, at least in some cases, the steps that the Ordnance Department had achieved in the direction of mechanized uniformity and interchangeability of parts paved the way toward the American system, and that private industry, instead of being the forerunner, followed in the path trodden by the Ordnance Department and its brotherhood of soldier-technologists.[39]

While this chapter may create the impression that by the late 1850s the Frankford Arsenal had abandoned all other activities to concentrate on mechanized production of small-arms ammunition, that is inaccurate. Much of the earlier work of the arsenal continued throughout this period. The arsenal received and repaired a large array of artillery pieces, small arms, and other military paraphernalia, and it still contracted for, received, inspected, and transshiped vast quantities of military goods purchased through the contract system. It also continued its relationship with the duPont Company, although instead of receiving the bulk of that company's production either the arsenal commander or his appointed representative visited the duPont mills, inspected the powder, and had it shipped from there directly to designated arsenals.

In spite of Major Hagner's industrial hustling, as the decade of the 1850s drew to an end the main activity of the Frankford Arsenal drifted from production to procurement. As if to avoid the growing sectional tension, the eyes of the nation and the War Department turned westward, and the arsenal focused on equipping companies of horse artillery with blankets, harnesses, and matériel rather than just on the production of percussion caps. Some historians have argued that this attention to westward expansion helps to explain why the War Department was caught ill prepared for the outbreak of the Civil War. The activities of the Frankford Arsenal in the late 1850s give credence to this argument. Now outfitted for large-quantity production of small-arms ammunition, it settled instead into a kind of industrial ennui.[40]

In the summer before the outbreak of the Civil War, political machinations once again shattered the calm of the arsenal. John B. Floyd, the Virginian serving as President James Buchanan's secretary of war, and Col. Henry Craig, the chief of ordnance, had engaged in a continuing dispute over appointments within the Ordnance Department and deployment of resources and personnel. In retaliation for Craig's obstinacy, Floyd ordered Maj. Peter Hagner removed from command of the Frankford Arsenal and transferred to the Ordnance Department's Siberia, Fort Leavenworth, Kansas. Floyd also appointed Josiah Gorgas, later chief of ordnance for the Confederacy, commander of the Frankford Arsenal.[41]

Peter Hagner had commanded the arsenal for nine years, longer than any of the fourteen officers who had preceded him. A preeminent soldier-technologist, he deserves credit for bringing the Frankford Arsenal into the industrial age. Perhaps as Floyd intended, Hagner's transfer left the Frankford Arsenal poorly prepared for the coming of the Civil War.

At the start of the Mexican war the Frankford Arsenal was a small but busy outpost of the Ordnance Department. The efforts of several enlightened secretaries of war, one persistent chief of ordnance, and a handful of increasingly professional officers explain why the arsenal not only continued to exist but flourished. However, the dozen years from the opening of the Mexican war wrought great changes. The demands of that war, the stubborn pursuit of greater uniformity in the production of arms and ammunition by the Ordnance

Department, and the skill and scientific knowledge of a small but professional officer corps all contributed to the evolution of the Frankford Arsenal from a rather obscure outpost to the center of mechanized production of small-arms ammunition. Besides these factors, there is another of equal importance. Peter Hagner, the soldier-technologist who presided over these changes, assembled a team of dedicated civilian craftsmen and soldiers of ordnance who possessed the skill to transform the idea of uniformity into "mechanical reality." Hagner succeeded not just by assembling this group of men and boys and paying them well, but by involving them in the decision-making process and respecting their judgment and initiative. After Hagner's departure, this group remained behind and stood ready when, during the Civil War, the arsenal took its second step toward industrialization—the evolution from small-scale mechanization to large-scale industrial production, and from a small post served mostly by enlisted soldiers to a large manufactory with hundreds of civilian employees.[42]

5

The Civil War and Its Aftermath, 1861–1867

This chapter focuses on the Frankford Arsenal during a relatively brief but important period, the Civil War years and immediately afterward. It took the Ordnance Department two years to meet adequately the insatiable demands for arms and ammunition. The need to fulfill those demands forced the department and the arsenal to take the final two steps toward completion of a wholly integrated industrial system, the erection of a large-scale industrial complex and the employment of a large number of civilian workers.

Mainly, this chapter describes the creation of those industrial sites, their operation, their wartime production, and the experience of that initial civilian work force. In the process several arguments introduced earlier will reappear. To be understood, industrialization must be seen as a complex system that involves more than simple mechanization of production. Workers who took employment in the factories and workshops accepted industrial labor as part of their cultural world as long as it allowed them to attain personal and familial goals. In this context, then, only threats to those goals and not vague social ideals created worker unrest. Although it was probably impossible for either side to be fully prepared for the coming of the Civil War, this chapter demonstrates that the Ordnance Department's apparent ill-preparedness was more illusory than real due to the existence of a skilled and dedicated cadre of professional officers, the soldier-technologists who commanded the Frankford Arsenal and who brought its industrialization to completion.

WAR, INTRIGUE, ACCUSATIONS, AND UNREST

The Civil War represented another step toward the nationalization of war begun by the French Revolution. Mammoth armies of citizen-soldiers replaced the small professional armies of earlier wars, and the new nation-states substituted their own ideologies for dynastic ambitions as the rationale for warfare. In turn these national imperatives called forth increasingly efficient weapons technology with which to annihilate those who threatened a nation's existence. The American Civil War fused these two elements, large armies and technological warfare, into an orgy of carnage unequaled in history.

Most histories have portrayed America as unprepared for the coming of this fratricidal conflict. In fact, in some respects both sides were well prepared for war. Psychologically, the sectional crisis of the 1850s had conditioned the people of the North and the South to see each other as enemies. Similarly, the rapid industrialization of the early nineteenth century had created the basic industries needed to produce the weapons and materials of war. Finally, both the Union and (to a lesser extent) the Confederacy counted among their ranks dedicated ordnance officers who possessed the skill and knowledge required to fashion the new industrial technology into the tools of war.

This group of soldier-technologists included Josiah Gorgas, whom Secretary Floyd appointed to replace Major Hagner as commander of the Frankford Arsenal. On the surface this was a logical and routine choice. Gorgas had graduated from West Point ranked sixth in his class, and he had performed with distinction as the ordnance officer on Winfield Scott's Veracruz expedition. In the intervening years Gorgas had also served at several major arsenals from Watervliet, New York, to Mount Vernon, Alabama. In addition to service at these diverse posts, Gorgas had other attributes that especially qualified him to supervise the munitions production at the arsenal. He had translated from the

German reports on experiments with cartridges that Alfred Mordecai had gathered in Europe in 1855 and 1856, and he had conducted experiments with both tin and brass cartridge cases. While stationed at Kennebec, Maine, Gorgas had solved that arsenal's flooding problem by construction of a retaining wall. The Frankford Arsenal had a similar problem with flooding, which earlier commanders had tried unsuccessfully to resolve.[1]

While the appointment of Josiah Gorgas seemed both normal and logical, in reality it bore the scars of the bureaucratic and political intrigue that so frequently marred the history of the Ordnance Department. Col. Henry Knox Craig, the chief of the Ordnance Department, and John B. Floyd, the secretary of war, had engaged in political sniping during Floyd's entire tenure. Floyd, a Virginian and a hard-line Southerner, endeavored to promote both Southern officers and the development of ordnance facilities in the South. Craig resisted behind the wall of bureaucratic obfuscation. Gorgas, no friend of Craig's, cultivated a close personal friendship with Floyd. When in June 1860 Craig attempted to transfer Gorgas to Fort Leavenworth, Floyd intervened, sent Major Hagner to Kansas, and appointed Gorgas as commander of the Frankford Arsenal in Hagner's place. Five months later Floyd again circumvented Craig's authority and appointed Gorgas to membership on the prestigious ordnance board.[2]

Josiah Gorgas's command of the arsenal proved short-lived. When the Confederate government invited him to become chief of ordnance of the Confederate army, Gorgas submitted his resignation to Colonel Craig and went south. Gorgas's departure severely aggravated the wartime difficulties at the arsenal. A junior officer without command experience, Lt. Thomas J. Treadwell, found himself in temporary command of an arsenal about to be overwhelmed as the Union government scrambled to arm for war. Furthermore, Gorgas's hasty flight southward left behind an aura of suspicion and clandestine intrigue that affected both officers and workers and disrupted production at the arsenal for almost a year.[3]

On April 17, 1861, the new commander, Capt. William Maynardier, wrote to Simon Cameron, the secretary of war, to inform him that a letter from a local inhabitant had appeared in a Philadelphia newspaper accusing Lieutenant Treadwell of shipping cannon primers to Gorgas, who now commanded the Confederate ordnance office. Within a month another letter—this time sent to Gen. George Ripley, now commanding the Federal ordnance office—tarred superintendent Robert T. Perkins with the same brush. Treadwell, an honorable officer, tried to interview his accuser but to no avail. In order to defend themselves, Lieutenant Treadwell and Robert Perkins were forced to spend valuable time making a "full statement" that included invoices and depositions. After expending much energy at such a critical time, Treadwell proved that he and Perkins were guilty only of the heinous crime of crating and shipping south the furniture and personal belongings of the Gorgas family. Vindicated, Treadwell escaped the episode. But Perkins's trouble had just begun. In a telegram on September 5 Lieutenant Treadwell informed General Ripley that

> Messrs. [Robert] Perkins and [Robert] Bolton [the master armorer] were arrested last evening by order of the Marshall [*sic*]. I have just seen the U.S. attorny [*sic*] who says they will be held for an investigation. The charges I understand are based upon the intercepted correspondence published yesterday.

Again at a critical time charges and arrests deprived the arsenal of its two most skilled workers and forced Treadwell to conduct an inquiry, after which he reported, "As far as I can judge the charges [are] . . . of the same nature as those heretofore investigated and answered to the satisfaction of the Department." Nine days later Treadwell again communicated with General Ripley, this time bearing better news:

> I have the honor to report that the investigation of the charges against Messrs. Perkins and Bolton . . . resulted in their unconditional acquittal. . . . I therefore . . . reinstate them.[4]

Throughout the early history of the Frankford Arsenal there is no doubt that some of the local residents regarded it as an acceptable and well-paying neighbor. The stonemason John Deal, the lumber merchant Frank Smedley, and the machine builder Barton Jenks, all engaged in business and other transactions with various arsenal commanders. John Deal's sister also served the arsenal as hospital matron until her death. Later, residents of the surrounding neighborhoods and their sons and daughters labored there in an atmosphere of mutual benefit. But at the same time other residents held the arsenal in much lower regard. This latest episode of hysterical accusations fit the pattern of hostility that at times disturbed the relationship between the post and its neighbors.

The Frankford Arsenal first appeared in the local records on the county tax list of 1817 as "U.S. Arsenal, 20 acres," assessed at $20,000. A tiny and barely viable property, it was assessed at nearly twice the value of even the most productive property in Oxford Township. Through the 1820s, 1830s, and 1840s, arsenal commanders employed their tactical training in a running battle with township officials over the tax assessment and over drainage of adjacent roads that dumped rainwater onto the grounds and undermined walls and fences. The city surveyor, a Frankford resident, tried to force the arsenal to tear down its front wall, claiming that it encroached on a public thoroughfare. This forced arsenal commanders to enlist the assistance of the United States attorney and to negotiate for land on the opposite side of the street at outrageous prices.

This animosity, particularly since it changed abruptly in the outpouring of patriotic fervor evoked by the Civil War, can be interpreted as evidence of the attitude of at least the public officials if not all of the residents of the surrounding community toward the arsenal. The intriguing problem of the true nature of local attitudes remains unsolved, however, since existing records give only the barest hint at the cause of the animosity. Was it Jacksonian antimilitarism, a vestige of Quaker pacifism, individual pettiness, natural resentment, or something else? Although the paucity of records precludes any certain conclusion, a common denominator united all of the arsenal's opponents. They all belonged to families that, during the eighteenth and early nineteenth centuries, had dominated Frankford and Oxford Township. Jesse Waln, the tax assessor, was the brother of Robert Waln, the bankrupt merchant. The family of David Scattergood, Treadwell's accuser, had been prominent in the local Quaker community. The families of Isaac Shallcross (the surveyor), Jacob Foulkrod (chief burgess of tiny Whitehall borough), and Jacob Pratt (owner of the adjacent property) had been the great landowners in earlier days before the intrusion of industrialization. Is it possible that their animosity toward the Frankford Arsenal represented an example of the focused status anxiety that Paul Johnson and other historians have identified in other early nineteenth-century elite groups?[5]

On November 8, 1861, Lieutenant Treadwell received a communication from General Ripley that exonerated him of all charges, commended him for his forbearance, and urged him to resume operations "with as little delay as possible." But for both Treadwell and the arsenal, the tribulations had just begun. The following month Treadwell reported an explosion in one of the smaller outbuildings in which workers fabricated cannon fuses. But Treadwell persisted, and by January 1862 he reported, "I am now ready to make up the fixed ammunition as fast as I can get the projectiles." However, in February an even greater disaster occurred: a fire severely damaged the building that housed the precious percussion cap machine. Miraculously, neither accident caused serious injury, and the arsenal work force labored mightily first to resume and then to increase production.[6]

These accidents and accusations, serious though they were, paled by comparison with later wartime problems, for throughout the Civil War petitions, demands, threats of strikes, and chronic unrest among workers plagued the industries of Frankford and other factories in the Philadelphia area. This continued unrest disrupted the arsenal more than any other event in the first sixty years of its existence, and it threatened to poison the relationship between the commander and the work force at a most critical moment—just as it became the employer of a large number of civilian workers.

After the assault on Fort Sumter the tempo of work at the arsenal quickened, and by the middle of the summer of 1861 it reached the frenzied pace that persisted until the end of the war. Workers labored both longer days and longer weeks, Saturday hours were extended, and Sunday, the Lord's day, became the government's day. Labor became continuous, but paydays became less regular. From the very outset of the war arsenal officers complained to the ordnance office, to the Treasury Department, and to anyone else who might listen. "I have no funds to pay the last months pay roll of hired men, most of whom really need the money due them," went one such missive. The ordnance office sympathized but did little other than to order, "in case of shortage of funds disbursing officer is to pay hired men before contractors." This gratuitous advice proved ineffectual, because payroll officers had no funds at all. To alleviate the crisis, the Treasury Department resorted to the ultimate placebo, paper money—or "Certificates of Indebtedness." Unfortunately, workers found that storeowners and other creditors would accept them only at discounted rates. This trapped workers between the inflated prices of wartime and deflated paper money.[7]

With obvious irritation, the commander defended the arsenal's pay scale. He polled manufacturers in the vicinity and replied to a complaining workman, "I am paying as much as they [other manufacturers] and rather more than Mr. Sellers. . . . Please communicate this to the workmen generally." To the wife of one of the arsenal's hired men he wrote, "I have no objection to your visiting the Arsenal but the employees at this Arsenal get high wages and there is no necessity for them asking for any more." Even Jabez Gill, inventor of an improved version of the cartridge machine and among the highest-paid workers, complained and received a sharp rebuff. "I am paying you as much as other manufacturers; I cannot therefore make any increase in your pay."[8]

While on one hand the commander defended the wages paid to workmen, he also acted as their advocate. Several letters that the arsenal commander either wrote or received are worth quoting at length, for they illuminate the plight of the workers and, most important, delineate their underlying cultural values. The first letter pleaded the cause of an enlisted laborer:

> Jacob Waltzhauer is a Corporal of Ordnance [and] was married about four years since and has two children.
>
> He purchased some time ago a house and a lot near the Arsenal + has paid $325 on it; a debt of $775 still remains unpaid. The high price of all the necessities of life, prevents him from saving anything from his pay, as

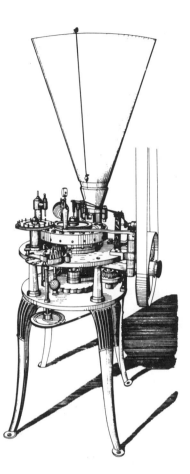

FIG. 15. Jabez Gill's cartridge-loading machine took mechanized munitions production one step further. It loaded both the charge of gunpowder and the conical lead (seen in a row on the middle level) into the preformed metallic, center-fired cartridge.

he has done heretofore, to meet the interest of the mortgage, and fears losing the sum he has already paid on the lot.[9]

On June 12, 1864, the commander received a letter from a group of laborers. It was written in pencil and signed by sixteen workers:

Owing to the enormous prices of articles of food and clothing and the consequent increase of living therefrom, and having discovered by the experience of the past two months, that recompense bestowed upon our manual labor is totally inadequate to our necessary wants, and to cover the expenses accruing therefrom.

We laborers together in this place unanimously petition that our daily wages be augmented to a sum commensurate with the conditions of the times, and adequate to our collective wants.[10]

A common grievance united the petitions of a skilled toolmaker, an enlisted corporal of ordnance, and a group of less-skilled laborers. All three wanted fair remuneration for their labor so they could provide their families with life's basic necessities, food, shelter, and clothing. None of the letters and petitions complained about long hours or loss of control over the conditions of the workplace. None advocated general economic or social change. Like other American workers in other times and places, they asked only to be paid fairly for work performed and to be treated reasonably. Although the prevailing economic distress strained the mutually beneficial relationship between the arsenal and its workers, the bonds did not snap. No work stoppage occurred. In the history of the Frankford Arsenal from its origin until its closure in this century, this is the only incident that even comes close to the description "labor unrest." It is the exception that proves the rule.

MAJOR THEODORE T. J. LAIDLEY, TRAGIC HERO

As the shelling of Fort Sumter reverberated across the nation in April 1861, Capt. William Maynardier, in very temporary command of the Frankford Arsenal, took stock of the ammunition on hand and found four million musket caps and two million pistol caps. He then decided to switch the pistol cap machine to the production of musket caps and reported to the ordnance office, "I expect to make at least 60,000 caps per day." By May the chief of ordnance summoned Maynardier to Washington, leaving the soon-beleaguered Lieutenant Treadwell in command. Treadwell informed the chief:

> The demand for P. Caps is greater than the product of the two machines and orders are still unfulfilled. I have made arrangements to run both machines at night until ½ past nine and one hour before the usual work hour, this will increase the daily product about 30,000.

By the following month the machine ran "day and night" and the operators increased production to 130,000 caps daily.[11]

When the fitful groping of Bull Run evolved into the titanic, long-drawn-out Civil War, the demand for ammunition quickly overwhelmed the production facilities of the Frankford Arsenal. In March 1862 Lieutenant Treadwell escaped to his new post in South Carolina, and Maj. Theodore Thaddeus Smolenski Laidley took his place. The Virginia-born Laidley had graduated from the United States Military Academy in 1842 ranked sixth in his class. He was a fit successor to Mordecai, Hagner, and Gorgas but perhaps excelled them as an innovative researcher and industrial builder.[12]

Laidley began confidently:

> I now make only 180,000 [caps] per day of 24 hours. In July I expect to make 7½ millions, and will add a new . . . machine every three weeks until I can make enough for the whole army.

After having mastered, at least temporarily, the munitions problem, Laidley turned his attention to other pressing issues. In particular he focused on the need to expand both production and maintenance facilities. Here he ran into the twin barrier of General Ripley's conservatism and Simon Cameron's corrupt incompetence. Ripley wrote: "this department is not prepared at this time to erect permanent buildings at the Frankford Arsenal." However, Laidley's persistence reaped rewards. First, he received permission to construct temporary buildings; then in July 1862 the secretary of war approved plans to erect a new blacksmith shop.[13]

In a modest way, the most significant expansion in the early history of the Frankford Arsenal had begun. The construction that started with the blacksmith shop and ended with the rolling mill in 1865 brought to fruition the industrial system of manufacturing initiated by Major Hagner with the opening of the percussion cap factory in 1854.

Even though the blacksmith shop was modest in size, it represented a major step in industrial construction. In light of the explosion that had severely damaged the percussion cap factory, Laidley designed a building that could be reconstructed rapidly in the event of a similar disaster. Other builders had used cast iron in buildings as early as 1852, but Laidley introduced wrought iron in place of cast iron. By using wrought iron columns and beams, Laidley made the walls superfluous. In the event of an explosion, walls would blow out and the roof would collapse; this would leave the columns and beams intact. Laidley equipped the new shop with two forges, piping, and a steam engine. He also moved to this shop other equipment that had been in the percussion cap factory, allowing for the installation of additional machines there.[14]

By 1863 several significant changes had taken place that promoted Laidley's building program. When the driving Edwin McMasters Stanton replaced the incompetent Simon Cameron as secretary of war, he forced General Ripley to resign as chief of ordnance. However, over Stanton's objection President Lincoln appointed George D. Ramsay, who had commanded the Frankford Arsenal both before and after the Mexican war, as the new Chief of Ordnance. Although Ramsay nominally held command, Stanton's protégé Capt. George T. Balch actually ran the ordnance office. By 1863 Jay Cooke's creative financing helped to provide Lincoln's administration with adequate funding. Collectively, these changes affected the Frankford Arsenal by making Laidley's construction projects possible.[15]

Initially, Laidley received permission to construct three additional buildings: a machine shop, a laboratory, and a rolling mill. The machine shop represented an especially important link in the evolution of the industrial system. First, Laidley took the new method of construction pioneered in the blacksmith shop and expanded it on a grander scale. Laidley designed the machine shop as a large (50 by 108 feet) two-story building with a central rear wing (54 by 40 feet) that made it T-shaped. As in the blacksmith shop, the columns, beams, and trusses supported the structure of the building. The brick walls, relieved of bearing weight, merely enhanced the aura of Victorian military monumentality. The structural columns, also designed by Laidley and manufactured by the Phoenix Iron Company, were particularly significant: they consisted of three curved sections bolted together. This represented a radically new use of iron in architecture.[16]

To fill this new building Major Laidley ordered a vast array of machine tools: presses, drills, lathes, planers, and a 150-horsepower Corliss steam engine to drive them all. Once again, faced with the foot-dragging of private manufacturers, the Ordnance Department had decided to manufacture its own machinery. Starting with the group of skilled craftsmen that Peter Hagner had already assembled, Major Laidley and his successor added other equally skilled machinists. Thus, in one shop was gathered a group of practical and innovative men whose sole function was designing, building, and maintaining machinery. Out of this research and development facility would come most of the innovations in munitions production for the remainder of the century. Historians of labor and industry have pointed to the importance of the new work force and new machinery in the evolution of industrialization. Architectural historians have noted the significance of Laidley's innovations in structural design. But what about the geography of the new workplace? What occurs when a group of very skilled workmen, adequately equipped and housed in a spacious facility, are permitted to interact? Does the new workplace affect industrialization as much as the new machinery?[17]

The next building held a special place in the heart of Theodore Laidley and his fellow technologists. The Ordnance Department intended it as a laboratory facility for the experiments in ballistics, chemistry, and metallurgy that the new technology demanded. Again Laidley designed a building in which wrought iron columns and beams supported the weight of the structure, still enclosed in

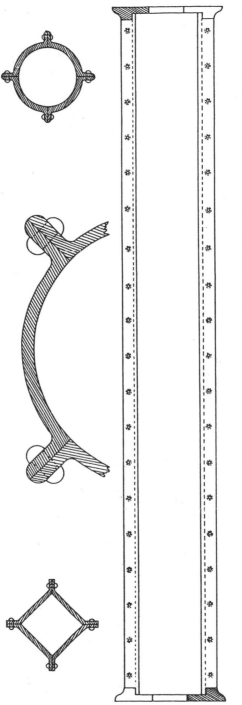

Fig. 16. This column, invented by Maj. Theodore T. S. Laidley and manufactured by the Phoenix Iron Company, both represented a revolutionary use of structural iron and cost Laidley his command of the Frankford Arsenal.

FIG. 17.   Building 40, the machine shop built by Major Laidley. Inside, George Wright, Jabez Gill, and other skilled craftsmen devised the cartridge-loading machines that secured the Frankford Arsenal's reputation.

massive brick walls. The main part of the building was 59 by 55 feet and two stories in height. Attached at either end of the main building were one-story wings, each 34 feet long. And arrayed in front were four subordinate outbuildings. Laidley's shopping list for the new laboratory included tools of experimental wizardry such as "3 new Vignotti machines with Rumpkoff coil, four densimeters and balances with weights."[18]

The final structure, the rolling mill, dwarfed the other buildings. For some reason that neither arsenal nor Ordnance Department records reveal, the department chose a Philadelphia architect, John Fraser, to design the two-story brick building that measured 225 by 50 feet. Although designed by Fraser, it—like the others—was constructed on Laidley's column and beam system.[19]

The monumental nature of the brick construction of these last three buildings, the employment of John Fraser, and the decision to hire large numbers of civilian laborers as arsenal employees all provide clues to the nature of the American

FIG. 18. Building 215, the central facade of the massive rolling mill, built in 1865 but never used for its intended purpose. The clock in the pediment is evidence of the imposition of order made necessary by the industrial system.

military mind in 1864. The Ordnance Department tried to use the size, shape, decoration, and arrangement of buildings to impose its authority, order, and power on civilian employees, enlisted men and officers alike. With initial reluctance, it hired civilians as its employees but used them in preference to employees controlled by outside contractors.

FIG. 19. This undated map shows the arsenal grounds sometime after the Civil War. In the upper left is the square parade ground surrounded by the commanding officer's quarters, two storehouses, and the original arsenal building. In the middle stand the T-shaped machine shop and the H-shaped percussion factory; the smaller buildings to the right are experimental laboratories. The upside-down U-shaped structure near Frankford Creek was meant to be the rolling mill.

Similarly, the Ordnance Department had come to a clear understanding of the industrial system and its relationship to that system, well in advance of private manufacturers. A letter from General Ramsay to Major Laidley clearly reveals that understanding:

> It is very desirable that this Department should be able to supply, from its own shops, all the Percussion Caps, Friction Primers and Copper Cartridge Cases required in the Service.
> With a view of accomplishing this object, as speedily as possible, you will take immediate measures to have plans prepared for shops, and all dependent detached buildings of sufficient to manufacture:
> 250,000,000 caps per annum

3,000,000 friction primers do do

20,000,000 copper cases do do

Starting from your present capacity as a basis you will divide all the operations connected with each of the above branches of manufacture in distinct groups, and from the means now at hand, make an accurate estimate of the number of each kind of machine necessary to bring your production up to desired standards. The capacity of each machine, the power required to run it, the superficial area which its attendent requires, and the amount of room in which each group of machines, required to accomplish one operation, can most advantageously be arranged, should all be given.

These details being ascertained, for each operation, under each branch of manufacture, as above indicated, shall be so combined together that, if possible, no part of the cap, primer, or copper case, from its first stage to the finished shall go over the same portion of the establishment twice.[20]

The Ordnance Department had come to understand that industrialization involved a complex system that stretched far beyond either simplistic mechanization or the wonderfully symbolic concept of interchangeable parts. It required management and organization of the diverse parts of that production, including the workers. Surprisingly (if one accepts the stereotype of the military mind as rigid) the department saw the system in its human dimension. Perhaps the experience of arsenal and armory commanders such as Peter Hagner, George Ramsay, and Theodore Laidley produced this attention to the human element in the industrial system. These commanders had involved creative and skilled workers in the decision-making process, paid them well, and allowed them both to create and to produce. They treated ordinary workers fairly and compassionately and remained cognizant of their human needs.

If a work of history needs heroes and tragic figures, in this study Maj. Theodore Laidley can fill both roles. In the spring and summer of 1864, while the Ordnance Department urged hasty completion of the works in progress, construction ground to a halt. Soon, unhappy letters passed between Laidley and the Phoenix Iron Company, letters reminiscent of the exchange a decade earlier between Major Hagner and the Remington Arms Company. When notified that the iron company was about to commence rolling beams, Laidley testily responded, "I understood you were rolling them. . . . " And when shipments failed to arrive he wrote, "You are going to ruin me . . . you have almost frightened me from the use of iron in buildings." Six weeks later he wrote again: "The sight of those unfinished walls . . . is humiliating in the extreme." Early in June Laidley received a terse note from the Phoenix Iron Company: "One of the 6″ rollers broke . . . will be several days before it is fixed . . . cannot roll 50′6″ beams." Finally, on August 11, 1864, the Phoenix Iron Company notified Laidley of the shipment of the long-awaited beams. On the same day Laidley received a letter from the ordnance office, Special Order 266. Laidley proved to be a prophet of his own fate; General Ramsay had removed him from command of the Frankford Arsenal.[21]

Through the end of 1864 and the beginning of 1865 Capt. Stephen Vincent Benet, the new arsenal commander, completed the elaborate building projects initiated by the ill-fated Laidley. In fairness, Benet labored under easier circumstances than Laidley. By the end of 1864 with the Civil War clearly drawing to a favorable conclusion, the War Department pressed its demands less insistently. But Benet, who would later hold the office of chief of ordnance from 1874 to 1891, seemed to possess some of the diplomacy lacking in Laidley. The remainder of the columns and beams from the Phoenix Iron Company arrived on schedule to hasten completion of the projects. (Laidley had perhaps made an error in ordering the ironwork from a company that also cast cannons for the Ordnance Department—and whose work Laidley occasionally rejected.)[22]

THE NEW WORK FORCE

By October 1864, 1,226 "hired men" labored at the Frankford Arsenal. Although the records identify these workers only as hired men, some—such as "Hugh Sullivan, foreman"—were or had been soldiers of ordnance. Benet employed several hundred of these workers as skilled craftsmen. Some of these carpenters, painters, and bricklayers labored on the building projects at wages that ranged from $2 to $3 per day; others such as blacksmiths and saddlers worked at the production of munitions and other items at wages from $1.75 to $3 per day.[23]

However, set above and around these skilled laborers, a number of workers filled the new jobs created by the industrial system. William K. Pigott, the son of the original William Pigott, and Robert T. Perkins, "master armorer" and "superintendent of works," stood at the top of the industrial hierarchy. Pigott received $4 a day and supervised eleven other clerks who were described as "working in the office." Perkins, the highest-paid employee, received $4.50 a day (during the year from October 1864 to September 1865 he averaged more than $100 per month) for supervising forty-one machinists as well as generally superintending the work of the arsenal.[24]

A large variety of other positions filled the niches of the industrial system

FIG. 20. This iron fence runs along both sides of the arsenal's main entrance. The decorated posts highlight the imposing military look its builders wanted to convey. The barbed wire is, of course, a more recent addition.

between the skilled craftsmen above and the laborers below. The range of these positions suggests the growing complexity of industry in the middle of the nineteenth century. They included the ominous "Timekeeper, J. Fleming, two dollars per day," "Draughtsmen," a "Sub-inspector of Powder," a "Superintending Fabrication of Primers," a "Superintendent of Laboratories," and several "foremen."[25]

More than a thousand employees held the most numerous and most common position—"laborer." This common identification, however, masked the variety of their labors. Receiving wages that varied widely, from \$.60 to \$2.40 per day, most of them, like the skilled craftsmen, worked at completing the four new buildings, especially the colossal rolling mill. But others labored at munitions production. They performed jobs such as "making percussion caps," "punching paper box bottoms, cleaning small arms, and packing, pinching + bundling cartridges." Unfortunately, since arsenal records identified employees only as "Gill, J. H." or "Walton, S.," they do not permit the reconstruction of the age or the gender of this new work force, the pinions upon which the new industrial system moved.[26]

With the completion of the building project and the end of the Civil War, the Frankford Arsenal fell upon hard times. The monumental rolling mill became a warehouse for unused weapons and ammunition, a place of storage rather than a place of production. By October 1866 the commander reduced the work force to 161 "hired men." Two of the machinists had gone, and those who remained found busywork such as "making hinges for [the] rolling mill door." Similarly, the laborers worked at "cleaning and packing small arms" or "breaking up small arms ammunition." But October 1866 marked the bottom of the decline. The records for the following month point to the arsenal's future as the focal point of small ammunition production for the United States Army for the next century. The period during and shortly after the Civil War marked the culmination of the evolution of small-arms ammunition with the arrival of the breech-loading, center-fired, self-contained cartridge. In November 1866 arsenal machinists began on "presses . . . center-fired cartridges . . . for Gat. gun." From January 1 to June 30, 1867, workers produced "40,350 cartridges, Gatling canister, 1 inch, 46,000 cartridges, Gatling balls, 1 inch, and 3,467,760 cartridges, rifle balls, .50 cal." And by November 1867 the number of "hired men" on the payroll had rebounded to 315 workers. The future of the arsenal was secure.[27]

This chapter described the Frankford Arsenal during the Civil War. In doing so it pointed to the two final steps in the creation of an industrial system of manufacture of small-arms ammunition: the construction of a true industrial complex of factories, machine shops, and laboratories; and the employment of a large civilian work force that ran the gamut from very skilled craftsmen to somewhat indifferently skilled common laborers. This chapter also demonstrated that by 1864 the officers of the Ordnance Department had come to understand and to implement the complexity of industrialization as a total system rather than simply the mechanization of production. At the end of the period the arsenal was in the hands of far-sighted, technologically oriented officers and capable civilian employees—and on the verge of a century of excellence in the research, development, and production of small-arms ammunition.

6

The Work Force of the Frankford Arsenal and the Industrial Community of Bridesburg, 1867–1870

By 1870 the Ordnance Department of the United States Army had created an industrial system for the manufacture of small-arms ammunition at the Frankford Arsenal. First the department had introduced mechanized production of pistol, musket, and rifle ammunition. Then in 1854 the arsenal commander, Maj. Peter Hagner, had added steampower. Finally, during the Civil War the department expanded the scope of this manufacture into a full-scale system that included specialized production, a large civilian work force, and extensive division of labor. After 1870 and into the twentieth century the industrial production at the arsenal continued to grow in size and technological complexity, but it retained all of the essential features that characterized it before 1870.

Most of the diverse civilian work force lived in an industrial community just beyond the gates of the arsenal. Their lives, their families, and their community became intertwined with industrialization. Strongly held cultural values led these workers to a pragmatic acceptance of industrial employment, even with its concomitant discipline and other personal restrictions—but only as long as it did not threaten other imperatives, especially home and family.

The records of the Frankford Arsenal do not make reconstruction of its work force easy. The post returns only identify workers by specific skill, daily wage, and total wages earned per month. For example, these records list "1 Machinist—5.00 per diem—115.00" or "70 laborers—1.00—1570.23." A second list, "Monthly Returns of Hired Men," identifies workers by last name and initials. It also includes the number of days worked each month, wages per day, and total wages earned.

Coyne, R. A.	26¾	1.00	26.75
Porter, J. G. (Foreman)	26½	2.50	65.62
Gill, R. M.	25¾	.90	23.17
Walton, S.	26¼	1.00	26.25

This second list does not identify specific workers with specific skills or jobs. The only clue to actual work appears in marginal notations beside the names of a group of workers—"crimping cases, punching cartg. cases" or "fabricating bullets, loading cartg . . . lubricating bullets." Neither of these two lists provides full names, ages, or gender of the worker.[1]

However, when these lists are compared to the Census of 1870 a more complete analysis becomes possible. The census gives detailed enumerations of households. It identifies the head of household and his or her sex, race, age, place of birth, and (for the first time) specific occupation. Additionally, it estimates the value of real and personal property belonging to the head of household. This census also identifies each other member of the household by first name, age, sex, race, and, where applicable, occupation. Unfortunately, it does not identify the specific place of residence of each enumerated household. Sometimes, apparently at the whim of the enumerator, it lists the street on which the family resided but no house number. However, most frequently the census simply lists households by ward or by precinct.[2]

Two factors preclude a completely substantial comparison of the arsenal records with the Census of 1870. First, because the last complete list of civilian employees dates from 1867, a three-year gap separates it from the census. Since the rate of mobility among arsenal workers appears to have been high, many of the 1867 workers may have left the area by 1870. Secondly, the Census of 1870 is flawed—so much so that a second enumeration was ordered for both the Twenty-Third and Twenty-Fifth wards, those in the immediate vicinity of the arsenal.

Table 1: All Frankford Arsenal workers, October 1867, distinguished by function, average daily and projected annual income.[1]

The total number of workers equals 211.			
Function	Number	Average Daily Wage	Projected Annual Income
clerk	1	4.00	1,200.00
master armorer	1	4.50	1,350.00
superintendent	1	4.25	1,275.00
machinist	38	3.76	1,128.00
foreman	4	2.44	731.25
laborer[2]	16	2.20	660.00
laborer[3]	150	1.73	519.60

[1]In the late nineteenth century, 643.00 was the annual income needed for a family of four.

[2]These sixteen laborers received a slightly higher daily wage for the more dangerous task of loading the powder charge into the metallic cartridge.

[3]Evidence suggests that a large number of this category of workers, perhaps the majority, may have been females or adolescents.

In October 1867, 211 civilian workers labored at the Frankford Arsenal. The Census of 1870 lists more than forty families in either the Twenty-Third or Twenty-Fifth Ward with one or more members who worked in the arsenal.

Among the male workers some familiar names appeared. William Pigott, one of the original civilian workers, had died in 1862, but he left behind at least one son holding a secure clerical position. The census identifies William K. Pigott as a thirty-seven-year-old white male. He headed a household that included three other persons with the same family name: Isabella, his wife; a two-year-old daughter, Stella; and twenty-two-year-old Hugh Pigott, certainly a younger brother. Although the Census of 1870 mentions neither real nor personal property, William Pigott had a monthly income of $108. This salary made him the fourth highest-paid arsenal employee.[3]

Robert Bolton, the fifty-year-old English-born "master armorer" who came from a substantial Frankford family, reported considerable wealth: $10,000 in personal wealth and $1,000 in real property. Even with such wealth Bolton's household still included a boarder, George Hober, a twenty-year-old hospital steward who was German by birth.[4]

In 1864 Jabez Gill had argued vociferously with Major Laidley, saying that he was underpaid and deserved greater remuneration. Apparently, Jabez Gill had resolved his differences with the arsenal commander, for in 1870 he, like Robert

Bolton, held the position of "Master Armorer, U.S. Arsenal." Although less wealthy than Robert Bolton, Jabez Gill owned $500 in personal property and $6,000 in real estate. The real estate included a large lot, a home, and an outbuilding (perhaps a stable) just down Bridge Street and across the creek from the gates of the arsenal. The prosperous John Lennig, son of the owner of the Tacony Chemical Works, lived across the street from Jabez Gill. Gill's income certainly allowed his family of six to live in considerable comfort.[5]

More than any other civilian, Robert Perkins had been responsible for the inception of mechanized production of small-arms ammunition at the Frankford Arsenal. By 1870 Perkins described himself to the census enumerator somewhat boastfully as a "retired gun manufacturer." Like Jabez Gill and Robert Bolton, Perkins, now sixty-three years old, possessed significant property: $3,500 in real estate and $5,000 in personal wealth.[6]

Clearly, some men prospered with the coming of the new industrial order. These men advanced either because they already possessed skills demanded by (or readily adapted to) the new technology, or because they developed the managerial and clerical talents needed in the increasing complexity of systematic industrialization. But what of others? What of the nonelite workers who labored in the shops and factories of the Frankford Arsenal?

Out of sixteen white, adult, male heads of household identified in the Census of 1870 as arsenal workers, thirteen possessed either real property, personal property, or both in amounts that varied greatly. While German-born, thirty-seven-year-old Michael Arnold reported $2,000 in real property and $400 in personal property, Alexander Gillett, a forty-nine-year-old native of New Jersey, listed no real property but $300 in personal wealth. Since Gillett's household included his wife, Anna, a widowed mother, and five children (three of whom were under thirteen), $300 in accumulated personal wealth seems like a substantial accomplishment.[7] Two of the three cases of arsenal employees who enumerated neither real nor personal property in the census can be explained easily.

Before the Civil War, John G. Porter, a blacksmith and a sergeant of ordnance, had been stationed at the Frankford Arsenal. In March 1861 he had transferred to a field regiment. In 1867 Porter, now a discharged civilian, held the position of foreman in the arsenal shop in which workers loaded cartridges with gunpowder. For this dangerous position that required unusually close and careful supervision of workers, John Porter received $65.62 in October 1867. By 1870 the thirty-three-year-old Porter had married and fathered two daughters, ages four and three. As a recently discharged soldier and a newly married father, John Porter would not have had time to accumulate any substantial real or personal property.

Patrick Toohy, also a former soldier of ordnance, was among the arsenal's lowest-paid adult male employees. Toohy, one of the arsenal's four watchmen, was a fifty-six-year-old native of Ireland and the father of a family that included his wife, Bridget, and six children, four of whom were old enough to be employed. Presumably, the wages of these four employed adult and adolescent children augmented Patrick Toohy's salary of approximately $65 per month, and this added income should have allowed the Toohy family a measure of security and comfort even without real or personal property.[8]

Although the paucity and unreliability of records prevents certitude, the evidence strongly suggests that both the elite and the ordinary workmen at the Frankford Arsenal found employment there satisfactory, and that in most cases their wages allowed them to accumulate real and personal property while supporting their families as well. But describing the adult male workers at the arsenal only begins a portrayal of its work force, for in addition to the men two other groups, women and adolescents, labored in its shops and factories.

FEMALE WORKERS

Again, the manuscript schedule of the Census of 1870 humanizes the age- and gender-neutral records of the Frankford Arsenal. Those records of "hired men" list "Walton, S." as having worked twenty-six and one-quarter days in October 1867 for a total of $26.25. The Census of 1870 transforms "Walton, S." into a human being. In 1870 Sallie Walton, twenty-nine years old, still worked in the arsenal. She performed mechanized work, tending a machine that loaded metallic cartridges with precise amounts of gunpowder. Sallie lived in a household with two middle-aged German immigrants of a different surname. Yet even this humanizing metamorphosis proves unsatisfactory, for we have no hint of why Sallie Walton chose industrial production as a source of livelihood. Perhaps she was the daughter of sixty-year-old Mary Walton, head of a household that included a thirty-three-year-old son and two younger children; maybe Sallie Walton was an independent-minded young woman who chose to live in a separate household and to work in a factory as part of some personal strategy. If that is so, Sallie Walton flies in the face of a considerable body of historical literature arguing that, in this age of Victorian domesticity, industrial work demeaned proper young women.[9] Sallie Walton had a lot of company. The Census of 1870 identifies nine other women who also labored in the shops and the factories of the Frankford Arsenal.

Harriet Winterbottom, an unmarried thirty-five-year-old woman who worked in the arsenal, lived in a house owned by a widow, Sarah Winterbottom—perhaps Harriet's sister-in-law. The household included one of Sarah's children, Robert, seventeen years old and "employed in industry." These incomes should have allowed the Winterbottom household a measure of economic security.[10]

Sallie Montgomery's case closely resembles that of Sallie Walton. A thirty-four-year-old woman, Montgomery lived in a home with several other unrelated people. Curiously, this property was owned by "the widow Winterbottom," and the household included nineteen-year-old Martha Winterbottom, Sarah Winterbottom's other child who also worked at the arsenal.[11]

One other case of a female worker fits into this pattern of female industrial employment. Census records identify Catherine Sullivan as a thirty-four-year-old, Irish-born head of household. Arsenal records describe her work as "cleaning accoutrements, equipment and harness." Catherine's widowhood must have been fairly recent, because her household included a one-year-old child, Juliann, as well as a seventy-year-old man, Patrick Sullivan. If Catherine's wages, approximately $330 per year, represented the only income for this family, it experienced hard circumstances.[12]

These cases demonstrate that some women accepted industrial labor in the shops, workrooms, and factories of the Frankford Arsenal as a way to earn a living. There is no evidence that they found such work to be either demeaning or degrading. Perhaps working together—even if it was under male supervision—allowed female workers to build a network based on common experience, which permitted them to share little joys and made life's tragedies more endurable.

ADOLESCENT WORKERS

As early as 1848 Lt. Andrew Dearborn, the arsenal commander, had defended to the Treasury Department the practice of employing the young sons of enlisted soldiers of ordnance. With the coming of large-scale industrialization this practice had evolved into an even more elaborate system. By 1870 arsenal and census records identify at least fourteen young men and women who were employed by the Ordnance Department at the Frankford Arsenal.[13]

Jabez Gill, Robert Bolton, and Robert Perkins were men of wealth, property, and position. All were highly skilled workers, and they were among the highest-paid workers at the arsenal. None needed the income from a working adolescent to sustain his family; yet all had at least one son who worked at the Frankford Arsenal.[14]

The case of Robert Bolton, Jr., might easily be dismissed as an example of nepotism. Like his father, twenty-three-year-old Robert Bolton, Jr., was a skilled and highly paid mechanist. He worked in the machine shop, probably under his father's supervision, and he ranked in the top ten percent in terms of salary. Yet both of Robert Bolton's two younger sons, Moses (nineteen years old) and Franklin (sixteen), worked in the arsenal as common laborers earning one dollar per day, nearly the lowest salary. Jabez Gill's seventeen-year-old son Ross, who worked in the percussion factory "fabricating and loading cartridges," earned even less—ninety cents per day—as did Robert Perkins's sixteen-year-old son William. Another skilled machinist, Julius Costa, also had a son, Charles (fifteen), who worked as a laborer. Census records describe Thomas Wallace as a "machinist general." Even though Wallace, like Costa, Gill, Perkins, and Bolton, possessed wealth in both real and personal property, both he and his fourteen-year-old daughter Kate worked at the arsenal.[15]

Neither nepotism nor the need for a child's earnings explains the employment of these young men and women in the industrial world of ammunition production. The cases of some other young men and young women help to build an alternative explanation.

By 1870 Pennsylvania law extended compulsory education to the age of fourteen. This meant that most young men and women finished their formal education at about eighth grade. Since high school education did not become commonplace until the very end of the nineteenth century, most fourteen-year-olds entered the work world of adults. In this case employment at the arsenal made perfect sense. Fathers could initiate sons into the world of industrial work under their own watchful eye. This strategy assured skilled and proud men like Jabez Gill and Robert Bolton that their sons would follow in their footsteps and become equally skilled and equally prosperous men, even if it meant that those sons had to start at the bottom as common laborers. For other families in this industrial community employment at the Frankford Arsenal served a double purpose. It meant that their children began in an occupation that promised prosperity, security, and prestige under the supervision of men such as Jabez Gill, John Porter, and the officers at the arsenal, who were respected members of the community. If the economic hard times that attended the end of the nineteenth century blighted that promise, it was no less alluring in 1870. Additionally, for some families the wages earned by sons and daughters made the difference between poverty and comfort.[16]

If Patrick Toohy earned sixty-five dollars a month as a watchman at the arsenal and his two sons each earned a dollar a day as laborers, then the total annual family income would have been just over $1,300, a fairly substantial amount. But if we include the wages of the two daughters, twenty-two-year-old Kate and fourteen-year-old Mary, who also worked in the arsenal, the total becomes close to $2,000. This amount greatly exceeds the $643 needed to maintain a minimum standard of living. Perhaps the Toohy family purchased carpets, a piano, and lace curtains that announced their prosperity to the outside world.

If the double households of the property-owning widow, Sarah Winterbottom, included four wage earners, two of whom (thirty-five-year-old Harriet and nineteen-year-old Martha) worked in the arsenal, their total wages should have been at least $1,200. This might explain how Sarah Winterbottom was able to own and maintain two dwellings and a property that exceeded six building lots in size.[17] From this perspective the employment of sons and daughters represents something other than exploitation. It can be interpreted as a strategy that allowed a family to maintain a decent standard of living and initiate young men and women into an industrial career that promised a prosperous future.[18]

By 1870 the Ordnance Department had established an elaborate industrial system for the manufacture of small-arms ammunition at the arsenal. In its shops and its factories labored a large, diversified, and mainly civilian work force. Both the arsenal and its workers shared a unifying experience: they both became part of an industrial community.[19]

THE INDUSTRIAL COMMUNITY

At the beginning of the nineteenth century the United States Army founded the Frankford Arsenal in a rural setting far from the congestion of urban Philadelphia. During most of its early history the inhabitants of Oxford Township had regarded the arsenal as an undesirable, alien intruder, but the coming of the Civil War and the subsequent employment of a large civilian work force changed all of that. From the end of the Civil War until the closing of the Frankford Arsenal in 1977, it and the community were intertwined. When they became supervising employers, once isolated officers were transformed into valued members of the community.

In 1870 the arsenal stood at the center of an already large but rapidly growing industrial community. Four communities abutted the arsenal. The village of Bridesburg, home to most of the arsenal's workers, lay to the south and east between Frankford Creek and the Delaware River. Meandering from the west, the creek, its banks lined with industries and businesses, acted as an industrial umbilical cord, and it attached the arsenal to the older village of Frankford. The borough of Whitehall, a tiny residential neighborhood and home of William Pigott and several other employees, adjoined the arsenal on its northwestern side, across Tacony Road. About a mile to the north at the end of Tacony Road sat the village of Tacony, soon to be the site of one of Philadelphia's greatest industries, Henry Disston's steel and saw works.

English-born Henry Disston had arrived in America in 1833 with his sister and father, only to be orphaned three days after his arrival. However, within seven years Disston had served as an apprentice wheelwright, and he had accumulated enough capital to begin his own saw-making business in Philadelphia. Disston built his industry with English steel and Sheffield steelworkers, another example of the transfer of English technology to America. By 1870 Disston had become alarmed by the urban contagion infecting his workers, and he decided to move his works to a 370-acre site he had purchased in Tacony. Beginning in 1872 he moved his operations there, shop by shop. A recent biography describes Disston as a benevolent paternalist. Like some other nineteenth-century industrialists, Disston, a Presbyterian, held his workers in tight control both in and out of the workplace. For his workers Disston built spacious and airy twin homes. He provided them with clean water, parks, and recreation facilities, but he also ran their lives. No taverns or liquor sales were permitted in his village. To avoid alarming workers, churches were forbidden to have bells in their steeples. Disston established a "beneficial" association to which workers contributed, but company officials disbursed the funds. Disston owned the village, its only bank, and its only newspaper. When some workers tried to organize to protest wage cuts in 1877 and 1885, Disston's son and heir fired them.[20]

The village of Bridesburg sits on a narrow and irregular triangle of land. On the east the Delaware River flows toward Philadelphia. On the west Frankford Creek separates Bridesburg from Frankford and the arsenal. The southern base of the triangle faces toward Philadelphia, about five miles away.

In 1870 in a fairly compressed area of less than a square mile lived more than 2,100 people, almost half of whom worked in one of a dozen factories. An 1875 map identifies about four hundred dwellings that were evenly divided among stone, brick, and frame construction and roughly equaled the number of households. In spite of the existence of several hotels and boardinghouses, few families shared dwellings. The same map identified the owner of each property. About a dozen persons, some of whose names appear in the earliest records of the village, owned about 40 percent of the properties. Individuals owned the remainder, but whether the owners also occupied the dwellings cannot be easily ascertained.[21]

The term "industrial community" certainly applies to nineteenth-century

Bridesburg. In the Census of 1870 only a handful of people still listed their occupation as "farmer" or "farm laborer." About fifty associated themselves with a commercial or service occupation such as "milk pedler" (*sic*), "domestic servant," or "working in soap store." The overwhelming majority identified themselves as industrial workers with descriptions such as "builder of cotton machinery," "working in arsenal," "working in chemical works," or "working in cotton mill."[22]

Three large factories—the Frankford Arsenal, the Bridesburg Manufacturing Company, and the Tacony Chemical Works—employed the majority of these workers, and each represented a different approach to industrialization.

In 1820 Alfred Jenks founded the Bridesburg Manufacturing Company to manufacture cotton textile machinery. Jenks, a native of Rhode Island, may have learned the design for such machinery from Samuel Slater. In 1830 the same Alfred Jenks invented a power loom for weaving checks. This loom was quickly introduced into the Kempton Mills in Manayunk. Sometime before the Civil War, Jenks and his son, Barton H. Jenks, established another factory, also called the Bridesburg Manufacturing Company, that wove cotton and woolen cloth. By 1870 the Bridesburg Manufacturing Company (the machine works) employed sixty-seven employees, all males over the age of sixteen. In that same year the firm was capitalized in excess of $1,000,000 and estimated its yearly production at $500,000. For the rest of the century Jenks's factory supplied machinery to textile mills throughout the country, but especially to the new industrialization of the South.[23]

In several important respects Alfred Jenks's approach to industrial management mirrored that of Henry Disston, the saw maker. Both men were active founders of Presbyterian churches, and both practiced an evangelical paternalism. They both found it necessary to control their workers in and out of the workplace. Jenks's work force was evenly divided between native Americans, mostly Pennsylvanians, and immigrants, principally from Scotland. The Census of 1870 lists thirty-five men as "cotton machine builders." These employees were also among the property owners in the village, and most, not surprisingly, also belonged to the Presbyterian church. Jenks had the reputation of being a stern employer who "somewhat controlled the morals of the town. . . . nobody who gambled or used profanity could work for him."[24]

The relationship between the various commanders of the Frankford Arsenal and their civilian employees stands in sharp contrast to the evangelical paternalism practiced by Henry Disston and Alfred Jenks. Nearly forty years records offer no evidence of any attempts by arsenal commanders to control the lives of their civilian workers beyond the gates of the post. Instead the records show that the various officers who commanded the arsenal between 1833 and 1870 used different management techniques. They tried to pay their workers well and frequently intervened with the Ordnance Department and other parts of the federal bureaucracy on behalf of their employees. Ample evidence has already been presented that demonstrates the extent to which arsenal commanders promoted technological creativity. The Ordnance Department even allowed employees to develop and patent several important technological innovations under their own names. Shops and offices were identified with employees, as in "Mr. Perkins' Shop" or "Mr. Pigott's Office." Clearly, the families in the vicinity trusted their sons and daughters to the supervision by the officers of the ordnance corps.

Perhaps different experiences account for the divergent approaches to the employer-employee relationship. Presumably, Henry Disston and Alfred Jenks came from the older, intimate, and familial school of the master-journeyman-apprentice relationship. The new industrial order, then, took these men outside the ken of their experience, and it forced them to grope for a substitute relationship. For both Henry Disston and Alfred Jenks evangelical Christianity, with its implication of moral preceptorship, offered an acceptable alternative to the vanished paternal authority of the master.

In contrast to Jenks and Disston, the officers of ordnance who commanded the arsenal drew on a different set of experiences. For centuries army officers had organized, in a military way, the labors of large numbers of men and boys. The soldier-technologists in command at the arsenal transferred these same techniques to a large civilian work force with relative ease, although one wonders if the same formula applied with equal ease to Sallie Walton, Catherine Sullivan, and the other female employees. However we account for the difference in the employer-employee relationships at the Frankford Arsenal and the factories of Alfred Jenks and Henry Disston, both of these models differed from the third large-scale employer in Bridesburg, Charles Lennig's Tacony Chemical Works.

In 1819 Nicholas Lennig, a German immigrant, had begun to manufacture chemicals in a factory in the Port Richmond section of Northern Liberties within Philadelphia County. The firm survived the depression of that year and the subsequent one in 1837, and by 1870 Lennig had moved to a twenty-seven-acre site in Bridesburg, just across Frankford Creek from the arsenal. Charles Lennig, the son of Nicholas Lennig, and Charles's cousin Frederick now operated the firm, sometimes referred to as C. F. Lennig. The Tacony Chemical Works distilled wood and produced metallic salts (especially sulfate of aluminum, used as a mordant in textile dying). Lennig's chemical works employed eighty-six men who, like Jenks's workers, were all over sixteen years of age. But in other respects the two groups of workers differed greatly.

FIG. 21.   Frankford Arsenal workers and machines for cartridge manufacture, Centennial Exposition, Philadelphia, 1876.

Almost without exception Lennig employed immigrants from southwestern Germany, from Baden and Württemberg. On the eve of the Franco-Prussian War this part of Germany had already experienced important social and economic changes. Centuries of partible inheritance had divided agricultural lands into plots too small for a family to farm, and bi-occupationalism had become the rule. If Lennig's immigrant workers resembled their nonemigrating countrymen, they had already experienced industrial work, especially in chemicals and metals, before their arrival in the Delaware Valley. This steady Germanic influx enhanced the already marked ethnic diversity of the village of Bridesburg and exemplifies a third approach to industrial supervision.[25]

Even amid an overwhelming industrial presence, Bridesburg retained some vestiges of a rural and more leisurely past. More than a mile of farmland and meadows separated Bridesburg from the rapidly growing and dividing metropolitan area of Philadelphia proper to the south. Summer cottages and fishing huts lined the banks of the Delaware River just to the south of Lennig's Tacony Chemical Works. During the summer months steam trolleys brought crowds of picnickers from the city. Watermen from the Delaware Bay purveyed their wares at the pier at the foot of Bridge Street. To the west of the pier the homes of John Lennig (son of Charles Lennig), Jabez Gill, and Alfred Jenks lined the same street before it crossed Frankford Creek to pass the imposing gates of the arsenal and rise toward the older industrial community of Frankford.[26]

COMMUNITY, WORKERS, AND VALUES

More than just a thoroughfare connected the newer industrial community of Bridesburg with the older and equally industrialized community of Frankford. Certain essential characteristics already noted in early nineteenth-century Frankford were mirrored in later nineteenth-century Bridesburg. In addition to industrialization, ethnic and religious pluralism and great geographic mobility marked both communities.

Like other nineteenth-century industrial communities, the town beyond the arsenal gates was marked by a rapid turnover of population. Most of the workers whose names appeared on the 1864 "return of hired men" had disappeared from a similar list by 1868. Most of those listed in 1868 cannot be found among the manuscript pages of the Census of 1870. Although this mobility is especially noticeable among propertyless laborers, it also applies to a lesser extent to people who owned property. Nearly half of those identified as property owners in the Census of 1870 do not appear on the 1875 map of Bridesburg.[27]

This constant human flux defies easy analysis. Geographic mobility does not translate into social mobility; moving out does not mean moving up, nor can the movement of people be automatically interpreted as occupational mobility. Sometimes families moved just around the block and moved as often as eight or nine times in ten years. In some urban communities April 1 was a kind of universal moving day. The only thing that one can readily conclude is that in late nineteenth-century Bridesburg like other urban communities people moved often and that rapid economic change impelled this movement, as though physically moving a person or a family reasserted a measure of control over one's life that industrialization denied. Yet even this defining characteristic was surpassed by another feature, ethnic and religious pluralism.

Within this compact community there existed five churches: All Saints, the Roman Catholic parish; the First Presbyterian church of Bridesburg, founded by

Alfred Jenks; a Methodist church; a German Reformed church; and a Dutch Zion congregation. An Episcopalian church and a Baptist church were located just outside the boundary of the village across Frankford Creek. However, even with such a large number of places to worship, records suggest that most of the people of Bridesburg, like other nineteenth-century Americans, remained unchurched.[28]

The diversity of Bridesburg can best be described by dividing the population into five groups: natives of Pennsylvania; non-Pennsylvanian Americans; and Irish, Scottish, and German immigrants.

The Census of 1870 identifies most of the inhabitants of Bridesburg as natives of Pennsylvania, although this is somewhat deceptive since a large number of these were Pennsylvania-born children of immigrant parents. This large number of Pennsylvanians offers evidence that rapid industrial growth continued to lure rural people to industrial communities. In addition, the industries of Bridesburg, and especially the shops of the Frankford Arsenal, attracted workers from other states up and down the Atlantic seaboard, from Massachusetts and Rhode Island to New Jersey and Maryland.

Three large immigrant groups joined these native-born Americans in the workplaces of Bridesburg. Charles Lennig's Tacony Chemical Works employed most of the considerable contingent of Germans; most of the Scots worked in the Bridesburg Manufacturing Company; and the equally numerous Irish tended to be scattered as "laborers," if they were male, and either mill workers or domestics, if they were female.[29]

Under some circumstances, especially economic hard times, the religious, ethnic, and cultural diversity present in Bridesburg may have served as a formula for tension and volatility, but there is no evidence of any such difficulty in the town except in one rather unexpected location: the Catholic church. When Bishop John Neumann established All Saints parish in 1860, he invited Redemptorist priests from Germany to staff the parish; mass was said in German. However, by 1870 Irish and native-born Americans made up the majority of the parish. Some tension continued between the German and non-German parts of the congregation for the remainder of the century.[30]

By 1870, then, Bridesburg possessed all of the characteristics of a late nineteenth-century American industrial community but none of the social conflict found elsewhere. It also lacked the spatial division by wealth already evident in nearby Frankford and metropolitan Philadelphia.

In the context of the rapid growth and ethnic, cultural, and religious diversity that marked Bridesburg, industrialism acted as a centrifugal force and a unifying catalyst rather than as a source of tension and division. This does not mean that all the people of Bridesburg saw industrialization from the same perspective and in the same way. The worlds of Alfred Jenks, Jabez Gill, Sallie Walton, and Patrick Toohy intersected only on a certain limited but important level. To Alfred Jenks industrialization meant wealth, prestige, power, and perhaps evangelical fulfillment. To Jabez Gill it meant comfort, recognition, independence, a future for his sons, and security for his family. To Sallie Walton industrial work may have allowed a measure of personal independence in a male world. To Patrick Toohy, an Irish immigrant, industry provided a degree of comfort and security for himself and his wife in advancing years and the promise of opportunity for his children. These four people and the other inhabitants of Bridesburg accepted industrialization even with its uncertainty and personal constraints, but they accepted it on their own terms and in light of their own values.

Conclusion

In 1816 the newly created Ordnance Department had no intention of engaging in the industrial production of munitions and other military supplies. It intended to have its needs met either by hand production by skilled soldiers of ordnance or by private manufacturers through the contract system. Changes in arms technology and the subsequent failure of private manufacturers to meet the increasingly rigorous tests devised by the young scientific-minded officers of the ordnance corps forced the department to move reluctantly in the direction of industrialization. This evolution began haltingly. In the beginning the department did not foresee and could not have foreseen the consequences of its decision to engage in mechanized production of ammunition for muskets, rifles, and other small arms, but within a thirty-year period from about 1835 to 1865 it came to see the manifold consequences of industrial production. Industrialization did not end at interchangeability of parts but was a complex system that included bureaucratic accountability, research, measurement of uniformity, large-scale specialized civilian labor, coherent distribution, and cost efficiency as well as mechanized production.

This evolution affected the Ordnance Department itself. It became large, expensive, bureaucratic, managerial, and powerfully influential. This evolution also affected the officers and enlisted soldiers of the ordnance corps. Once isolated, military men now became integral parts of the community in which the department's installations were located.

The creation of this complex industrial system also changed work and affected workers. Work became faster, more complicated, and more specialized. New jobs that demanded new skills appeared in both production and management, but old skills still retained marketable value. Clerks, timekeepers, foremen, supervisors, draftsmen, chemists, and machinists entered the work force. This work force became both specialized and elongated. Those at the top who possessed the most valuable skills prospered; those at the bottom labored for insufficient wages. However, this division of labor did not produce a permanent proletariat at the Frankford Arsenal.

The laborers, the lowest-paid workers at the arsenal, consisted mainly of two groups of temporary employees, women and adolescents. For some women employment at the arsenal may have meant personal independence. For several widows it meant the survival of their families. For young women and some young men, the wages from arsenal employment made the difference between subsistence and comfort for their families. For all of the young men, employment at the arsenal represented initiation into the work world of adulthood with the promise of future security.

Given the nature of the work force at the arsenal and the high degree of mobility that characterized it, the absence of any evidence of class consciousness within it is hardly surprising. The only example of unrest among the workers came during the Civil War, and even then workers reacted to economic distress caused by wartime inflation and currency instability rather than to work conditions. They exhibited no broad social grievances, and they complained only when the perceived needs of their families were threatened.

These findings lead to some larger suggestions. A fair number of American historians have used a Marxist model to investigate the impact of industrialization on American workers. This model assumes that industrialization is oppressive because it degrades skilled workers and that the protagonists in the ensuing struggle are large, cohesive groups: employers, middle-class bourgeoisie, and workers. It usually focuses on workers and their reaction to changes in work rather than on the actual work they perform. Finally, if historians employing the Marxist model do discover any class consciousness, they are forced to search for the causes of the ephemeral existence and rapid dissipation of that illusory mind-set.

The evolution of the industrial system for the production of small-arms ammunition at the Frankford Arsenal suggests a possible alternative model focused on technology. Examining technology allows us to study workers, work, and changes in work. It enables us to see workers reacting and adjusting to industrialization both positively and negatively. It shows us workers pitting their values and cultural imperatives against the abrasive influence of industrialization. Lastly, it permits us to see industrial communities as dynamic places in which individuals, ethnic and other cultural groups, families, employers, and employees constantly readjust and renegotiate the conditions of their lives, with changing industrial technology as a focal point.

The findings of this study concur with other recent works that suggest that terms such as "preindustrial" and "Industrial Revolution" do not help but rather mislead us. Many nineteenth-century Americans and many immigrants from both Great Britain and continental Europe were bi-occupational and familiar with early but fairly sophisticated technology. For these workers industrial employment was neither revolutionary nor threatening. They already possessed traits—mobility, adaptability, and market-oriented economic values—that permitted relatively easy initiation into full industrial employment. In this context industrialization represented not upheaval and disruption but a continuum of change that required further readjustment from people coping with a constantly changing world.

When the Frankford Arsenal closed more than two thousand jobs vanished. Buildings stood silent and empty. Machinery was removed and left to rust in open fields. Some people moved in search of new employment, but most stayed. Once again workers readjusted to the changing conditions of the industrial order.

Epilogue

The end of the Civil War and the exhausted silence that replaced the clash of arms caught the Ordnance Department and the Frankford Arsenal as unprepared as had the beginning of the conflict. Work in progress ground to a halt. The huge, recently completed rolling mill stood empty; nothing would ever be rolled there. Soon this monumental building became a warehouse in which the Ordnance Department stored arms and ammunition collected from disbanding armies. The "hurry up and stop" cycle that marks the whole history of the arsenal skidded to a halt once more. The arsenal commander found busywork for the remnants of a rapidly reduced work force. Skilled machinists fashioned hinges instead of machines. Laborers "broke up" ammunition and prepared it for storage rather than fabricating it.

Yet in spite of the appearance of dissolution, the future of the arsenal had been secured. Within three years the shops of the arsenal began developing, testing, and producing modern small-arms ammunition, a center-fired metallic cartridge containing both a charge of gunpowder and a conical bullet, in increasingly diverse types and calibers. From 1868 to the middle of World War II the Frankford Arsenal developed and fabricated virtually every type of small-arms ammunition used by the United States military.

As the nineteenth century drew to a close, the Ordnance Department expanded the operations of the arsenal to include several additional functions. Under the early soldier-technologists Alfred Mordecai, Peter V. Hagner, and T.T.S. Laidley, the arsenal established a reputation for scientific experimentation. By the beginning of the twentieth century, thanks to a continuing phalanx of scientific-minded officers and civilians—Lt. Beverly Dunn, Capt. John Pitman, Lt. Walker

Benet, Lt. Frank Baker, the British chemist and fellow of the Royal Academy of Science William D. Williams, and L. D. Boody — the laboratory research rivaled and even surpassed munitions production as the main business of the Frankford Arsenal.

During the Spanish-American War Captain Pitman began developing smokeless gunpowder. After the war William Williams built on this research to make smokeless gunpowder burn progressively. Lieutenant Dunn invented a new explosive, "dunnite" or ammonium picrate, which was of value to naval ordnance because of its moisture-resistant properties. By 1900, in conjunction with the duPont Company, the Frankford Arsenal had become the center for explosives testing in the United States.

By the end of the nineteenth century the evolution of field artillery was keeping pace with that of small arms, shoulder arms, and munitions. During the Civil War a battery officer using field glasses could direct the firing of muzzle-loading cannons by sight. With the improvements in the design, construction, and range of field artillery by the Spanish-American War, visual sighting was no longer possible. Accuracy in directing the fire of technologically sophisticated weaponry required equally sophisticated instrumentation. Once again, as was the case with the Remington contract in the 1850s, the Ordnance Department's requirements for accuracy exceeded the capacity of private industry. To meet this need the Ordnance Department in 1901 added a third division, the Instrument Department, housed in its own specially constructed building, to join munitions production and munitions research as the corporate entity of the Frankford Arsenal in the twentieth century. This instrument division not only developed and manufactured artillery sights, but also the instruments and gauges used to control the quality of their production as well as their calibration. One of the shops within the department, the optical shop, attained special prominence. By the 1930s the optical shop developed the capacity to grind lenses as large as ten inches in diameter that surpassed even the rigorous standards set by the United States Bureau of Standards.

For the Frankford Arsenal the outbreak of World War I repeated the pattern already established by earlier conflicts. First, the Ordnance Department purchased additional land, the twenty-eight-acre farm along the arsenal's northern and eastern boundary that was owned by the local industrialist John Lennig. This purchase both increased the size of the compound and gave it direct access to the Delaware River rather than just the tributary Frankford Creek. New building commenced, and production accelerated. During the Spanish-American War the arsenal had manufactured 37 million rounds of small-arms ammunition annually.

By the end of World War I this production reached 60 million rounds per year, and fifty-five buildings crowded the arsenal grounds.

For a final time, in spite of the escalating tension in Europe and the Pacific, the onset of World War II found the United States Army ill prepared. In 1940 the Frankford Arsenal produced all of the military ammunition in the United States. Faced with this inadequacy, the Ordnance Department repeated the now familiar pattern. Production increased; in the month of October 1942 alone the arsenal produced and shipped almost six thousand tons of various types and sizes of ammunition. The department added more land that extended the arsenal's northern boundary again: it purchased the 8.8 acres belonging to the Fitler Cordage Works. New buildings mushroomed. By the end of World War II the arsenal's confines burgeoned with more than two hundred buildings housing more than six thousand workers. But this time the War Department's plan featured a significant difference: the Government Owned–Contractor Operated Plan, or "Go-co."

This program dispersed production knowledge. Representatives from private industries—Eastman-Kodak, Remington, Winchester, Westinghouse, Bendix, Hamilton, and others—flocked to the arsenal's shops and factories, where they were trained by arsenal personnel. They departed laden with drawings, blueprints, specifications, gauges, and other technical instruments and data. By increasing industrial production to win World War II, the program was spectacularly successful; but Go-co proved to be the death knell of the Frankford Arsenal. By the conclusion of the war these private industries had stripped the arsenal of both its knowledge and its production facilities. Even with brief revivals during the Korean conflict and the Vietnam War, the arsenal again fell victim to political machinations. In 1976, in spite of election promises by candidates and government officials, the Defense Department announced the closing of the Frankford Arsenal. After removing its remaining research and development facilities (the vaunted optical shop was transferred from its specially constructed building to a quonset hut at the Picatinny Arsenal in Dover, New Jersey, while other work went to the Rock Island Arsenal in Illinois), the federal government handed the property to the city of Philadelphia. In 1981 the city sold the property to a private developer, Mark Hankin, who now operates the property as the Arsenal Business Center.

Today, under Mr. Hankin's proprietorship, the older buildings have been listed on the national register of historic places. They and the grounds have been carefully maintained. High-tech firms lease some of the buildings; a few of the older buildings, including the 1865 rolling mill and the percussion cap factory, remain empty. They seem to stand as patient, silent sentinels—as though they have been through these hard times before, and await the return of the sounds of machines and the voices of the workers.

Perhaps we can listen to the voices of some of those workers:

Arthur Hewitt:
  position: engineer and manager in the Instrument Department
  dates of employment: 1940–70
Question: What was it like to work at the Frankford Arsenal?
Answer: A tremendous place to work . . . good people to work with. . . .
Q: What do you remember best about the arsenal?
A: I think of the quality . . . the quality of the product . . . the ability of the skilled workers.
Q: Can you think of any specific examples of the quality?
A: Once, after we had ground a new ten-inch lens, we sent it to the U.S. Bureau of Standards to check its quality. I couldn't read all of the technical language of the report that they sent back, but the last two sentences were clear enough. They said it was the best optical flat they had ever seen. It was better than their own standard model.

Helen Hewitt:
  position: first small-arms ammunition, then procurement
  dates of employment: 1938–45
Q: What was it like to work at the Frankford Arsenal?
A: Wonderful . . . the best job I ever had . . . but hard work.
Q: What kind of work did you do in small-arms ammunition?
A: There were about fifty women working at machines loading cartridges. Several of us worked as inspectors. We would use micrometers and gauges to check the quality of the cartridges.
Q: How did you move from small-arms munitions to procurement?
A: One of the bosses saw the notice and recommended that I apply for the promotion. Working in procurement was really interesting. Since I had worked in ammunition production, I was familiar with the product and I was able to converse with the engineers and technicians when they called for specifications.

Ed Carroll:
  position: started as a apprentice toolmaker, became the director of manufacturing for ammunition, small arms, fuses, and fire control
  dates of employment: 1939–73

Q: What was it like to work at the Frankford Arsenal?

A: It was the greatest experience I ever had. There was great opportunity for career advancement. The toolmaking program was unmatched. The dedication of the people in the organization was unmatched by private industry.

Q: Can you give any examples of the excellence of the work?

A: Yes. Quick response. In time of war or other possible military need the Frankford Arsenal had the ability to, say within nine months, begin to produce any item before private industry could begin to mobilize.

Kathleen Liggins:
> position: inspector, industrial x-ray of munition
> dates of employment: 1951–76

Q: What was it like to work at the Frankford Arsenal?

A: A nice place to work . . . had control . . . a challenge . . . different work.

Q: What exactly was industrial x-ray?

A: Instead of using a water test to test cartridges for flaws a scientist at the arsenal invented a method of using radioactive krypton gas. This method didn't damage the cartridges like the old water test did. The work was hazardous, but I enjoyed the job. I was the only woman licensed by the Atomic Energy Commission to operate the machine.

Q: What was it like to be a black woman working at the Frankford Arsenal?

A: It was a good place to work. I felt like I was respected for the work that I did. Once I refused to certify a catapult [part of the mechanism to eject a pilot from a jet aircraft]. The boss gave me a hard time but I still refused. Later, he came back and apologized.

Q: Do you remember any specific incidents?

A: Once I chased a colonel from the room where I was microfilming some blueprints. You needed a security clearance to be in the room. He said, "Do you know who I am?" I said, "I don't care who you are. You can't come in here unless you have a 'need to know' [security clearance]." I know he complained to my boss, but I never was told about it.

Clare Readinger:
> position: secretary to deputy chief of research (Pitman-Dunn Laboratory)
> dates of employment: 1955–62

Q: What was it like to work at the Frankford Arsenal?

A: I enjoyed my job . . . it was a good place to work. . . . I was paid well. . . . There was never any tension or altercations, not even with a supervisor . . . they were always willing to help with personal problems.

Q: Do you remember any specific incidents?
A: The arsenal was very active in community events like Toys for Tots. . . . I once got a commendation and an award of $150 as part of an incentive program . . . I was recommended by a supervisor for good work.

Consistently and without exception the voices of the twentieth-century workers tell us what the nineteenth-century records only hinted at. The Frankford Arsenal was considered a great place to work.

Appendix

COMMANDERS OF THE FRANKFORD ARSENAL, 1816–1869

1. Captain Joseph H. Rees, August 1816–February 1821
2. Lieutenant Martin Thomas, Jr., February 1821–September 1824
3. Lieutenant Thomas J. Baird, September 1824–November 1827
4. Lieutenant E. M. Eakin, November 1827–April 1828
5. Lieutenant Charles Mellon, April 1828–December 1830
6. Major Joseph B. Walbach, December 1830–October 1832
7. Major William J. Worth, October 1832–January 1835
8. Captain Alfred Mordecai, January 1835–October 1838
9. Captain George D. Ramsay, October 1838–July 1845
10. Major Henry Knox Craig, July 1845–October 1845
11. Lieutenant Andrew H. Dearborn, October 1845–March 1848
12. Lieutenant L.A.B. Walbach, March 1848–October 1848
13. Major George D. Ramsay, October 1848–September 1851
14. Major Peter V. Hagner, September 1851–July 1860
15. Captain Josiah Gorgas, July 1860–March 1861
16. Captain William Maynardier, April 1861–March 1862
17. Major Theodore T. S. Laidley, March 1862–August 1864
18. Captain Stephen V. Benet, August 1864–October 1869

Notes

INTRODUCTION

1. An example of the economic determinism model in armament production is Felicia Johnson Deyrup, *Arms Makers of the Connecticut Valley: A Regional Study of Economic Development of the Small Arms Industry, 1798–1870,* Smith College Studies in History, vol. 33 (Northampton, Mass., 1948). An example of technological exclusivity in the explosive industry is Arthur Pine VanGelder and Hugo Schlatter, *History of the Explosive Industry* (New York: Macmillan, 1927).

Among the more recent works that emphasize the context in which technological developments take place are: Judith McGaw, *Most Wonderful Machine: Mechanization and Social Change in Berkshire Paper-Making, 1801–1885* (Princeton: Princeton University Press, 1987), introduction; Merrit Roe Smith, *Harpers Ferry Armory and the New Technology: The Challenge of Change* (Ithaca: Cornell University Press, 1977); and Robert B. Gordon, "Who Turned the Mechanical Ideal into Mechanical Reality?" *Technology and Culture* 29 (October 1988), 744–778.

2. Among the studies produced by the new social history that established the methodology and the model for subsequent studies of early American communities are: John Demos, *A Little Commonwealth: Family Life in Plymouth Colony* (New York: Oxford University Press, 1970); Philip J. Greven, *Four Generations: Population, Land, and Family in Colonial Andover, Massachusetts* (Ithaca: Cornell University Press, 1970); Kenneth A. Lockridge, *A New England Town: The First Hundred Years, Dedham, Massachusetts, 1636–1736* (New York: W. W. Norton, 1970); and Michael Zuckerman, *Peaceful Kingdoms: New England Towns in the Eighteenth Century* (New York: W. W. Norton, 1970).

A wealth of sources make this socioeconomic reconstruction possible. Several early nineteenth-century tax lists for Oxford Township, the township in Philadelphia County in which Frankford was located, are housed in the collection of either the Archives of the City of Philadelphia or the Pennsylvania Historical Society. These tax lists identify the members of the household as well as the assessed value of the property and the type of structures built on the property. When this information is supplemented by information from manuscript census returns, church membership lists, and other sources, a description of the village of Frankford as it evolved from a manufacturing village to an industrial community in the first half of the nineteenth century becomes possible.

3. A number of sources make this history of the Frankford Arsenal feasible. The Philadelphia branch of the National Archives holds an extensive collection of arsenal documents. This collection included sixteen bound volumes of copies of letters and reports written by various arsenal commanders,

eight boxes of letters received from the office of the chief of ordnance, and fourteen boxes of letters received from other sources, along with several volumes related to hired civilian workers.

4. Russell F. Weigley, *History of the United States Army,* enlarged edition (Bloomington: Indiana University Press, 1967), chapters 7 and 8.

5. Ibid.

6. Within the last thirty years a number of historians, both European and American, have examined the impact of industrialization on skilled workers, but since these historians employ a Marxian model they emphasize the exploitation of these workers rather than the actual work that these workers performed. Among the most important of these studies are: Eric J. Hobsbawn, *Labouring Men: Studies in the History of Labour* (London: Weidenfeld & Nicolson, 1968); George Rude', *The Crowd in History: A Study of Popular Disturbances in France and England, 1730–1848,* revised edition (London: Lawrence & Wishart, 1981); E. P. Thompson, *The Making of the English Working Class,* Vintage edition (New York: Alfred A. Knopf and Random House, 1966); Alan Dawley, *Class and Community: The Industrial Revolution in Lynn* (Cambridge, Mass.: Harvard University Press, 1976); and Bruce Laurie, *The Working People of Philadelphia, 1800–1850* (Philadelphia: Temple University Press, 1980).

7. Donald R. Hoke, "Ingenious Yankees: The Rise of the American System of Manufacture in the Private Sector" (Ph.D. dissertation, University of Wisconsin–Madison, 1984); and Merrit Roe Smith, "Army Ordnance and the 'American System' of Manufacture, 1815–1860," in Merrit Roe Smith, ed., *Military Enterprise and Technological Change: Perspectives on the American Enterprise* (Cambridge: MIT Press, 1985), 39–58.

8. Tamara K. Hareven, *Family Time and Industrial Time: The Relationship Between the Family and Work in a New England Industrial Community* (New York: Cambridge University Press, 1981), especially chapters 3, 5, and 8.

CHAPTER 1

1. Stephanie G. Wolf, *Urban Village, Population, Community, and Family Structure in Germantown, Pennsylvania, 1683–1800* (Princeton: Princeton University Press, 1976), draws a sharp contrast between the "urban village" of Germantown and the studies of New England communities (cited in the introduction to this book) that too frequently have been accepted as the American prototype. Wolf portrays Germantown as urban rather than rural in its pattern of settlement, as commercial and industrial rather than agricultural in its economy, and as associational rather than communal in its social and political structure. Furthermore, she describes the population as heterogeneous and mobile rather than stable and uniform, and its families as smaller and nuclear rather than extended. Finally, Wolf points out that the pattern both in Germantown and in the larger Middle Atlantic region more closely resembles future nineteenth-century conditions than does aberrant New England.

2. Attempts at the reconstruction of the earliest history of Frankford present a historian with a formidable challenge. Records are incomplete and sketchy. Personal recollections, letters, diaries, and newspapers are apparently nonexistent, and facts are interwoven with local legends. Most of the evidence for the earliest period consists of land surveys, petitions, county tax records, a federal dwelling assessment list, and the first several United States censuses.

Until the consolidation of the city of Philadelphia in 1854, Frankford was part of Oxford Township in the county of Philadelphia. The Philadelphia City Archives and the Historical Society of Pennsylvania hold some of these records from 1800 to 1817. The records identify, by name, owners of assessed property, their occupation, the property, and its assessed value.

In 1798 the U.S. Congress proposed a direct tax on dwellings as a source of income for the federal government. Even though Congress never enacted the tax, in May 1799 appointed officials surveyed prospective taxable dwellings. These lists identify the occupant of the dwelling, the owner of the dwelling (many were leased or rented), its location, the adjoining neighbors, the dwelling itself ("32 × 24, 2-story stone"), and its value.

The Census of 1810 lists each head of household by name and enumerates the number of persons in that household by sex, age, and race, but it does not identify them individually by name. The Census of 1820 provides even more useful information. In addition to the name of the head of household, the total number of inhabitants in that household, and their sex, race, and age, it also categorizes the occupation of the employed in the household (agriculture, commerce, or manufacture) and identifies "foreigners not naturalized." The 1830 and 1840 censuses for Oxford Township are illegible and useless. The Census of 1850, the last used here, provides information in thirteen different categories, including (for the first time) the name, age, sex, and race of each member of the household along with specific occupations, values of real property, and places of birth. In addition to governmental records, records of some churches also help to identify the religious affiliation of many Frankford inhabitants.

3. Early Land Surveys in Pennsylvania and Delaware, in the Philadelphia County Papers, Historical Society of Pennsylvania (hereafter HSP); Norris MSS, Philadelphia and Other Counties (1761), HSP; Logan Papers, Report to the Provincial Council on the Frankford Road, Sept. 21, 1701, HSP; Logan Papers, Frankford Road Surveys by Isaac Norris, March 1729/30, HSP; Misc. Papers of Phila. County, 1671–1738, HSP; and Charles H. Duffield, "The Old Swedes Mill: Its Water Rights and Surrounding Industries," June 1907. (This is an unpublished, typewritten manuscript held by HSP. The Duffield family purchased the mill site in 1800, and Charles Duffield possessed copies of the original deeds.)

For information on the Society of Free Traders see Document 34, "Charter for Free Society of Traders," in Jean Soderlund, ed., *William Penn and the Founding of Pennsylvania, 1680–1684: A Documentary History* (Philadelphia: University of Pennsylvania Press and the Historical Society of Pennsylvania, 1983), 147–152, especially the maps on 214 and 337.

4. Duffield, "The Old Swedes Mill," details Miller's purchase of the gristmill. "Historical report" in the files of the Philadelphia Historical Commission (hereafter Hist. Comm.) contains information about the Neffs and several other early industries. The information on the powder mill is from VanGelder and Schlatter, *History of the Explosive Industry,* 62–80.

5. Direct Tax of 1798, a microfilm roll, no. 5, M372 in the holdings of the Philadelphia branch of the National Archives; and the Philadelphia County Tax List 1800, Oxford Township, a bound manuscript volume in the collection of HSP.

6. Ibid.

7. Ibid.

8. In March 1800, in recognition of the existence of the village of Frankford as a separate entity, the Pennsylvania legislature had chartered the borough of Frankford, which continued in place until 1872. See Frankford Borough Ordnances, 1840–1846, City Archives.

9. In her study of Germantown, Stephanie Wolf found an evolution toward a smaller nuclear family. In Frankford older patterns of larger families persisted into the middle of the nineteenth century.

10. Census 1810, Microfilm, no. 56, M252, National Archives; and County Tax List, Oxford Township, 1810, City Archives.

11. Census 1820, Microfilm, no. 33, M109, National Archives; County Tax List, Oxford Township, 1810, City Archives; William B. Dixon, "Frankford's Early Industrial History," *The Pamphlet for the Historical Society of Frankford* 2, no. 3 (1911), 50–59, in HSP; George Castor Martin, "Samuel Martin, Proprietor of the Textile Mill in Frankford," papers read before the Historical Society of Frankford, a collection of unpublished material held by the Philadelphia Historical Commission.

The reason why Frankford seemed destined to become the umbrella capital of America remains a mystery.

12. Mary McConaghy, "The Whitaker Mill, 1813–1843: A Case Study of Workers, Technology,

and Community in Early Industrial Philadelphia," *Pennsylvania History* 51 (January 1984), 30–53; Ms. McConaghy based this study largely on company records preserved by the Eleutherian Mills Museum.

13. For the search and the decision to purchase the arsenal site, see: Col. William Linnard to Col. George Bomford, Sept. 19, 1815, Orders Received, June 1815–November 1818, Record Group 74, Entry 43, National Archives; and Capt. Joseph H. Rees to Col. George Bomford, Letterbook as cited in *Historical and Archaeological Survey of the Frankford Arsenal, Philadelphia, Pennsylvania,* prepared for the Department of the Army, Baltimore District Corp of Engineers, Baltimore, Maryland, by John Milnor Associates, May 1979, 2 (hereafter *Historical*).

14. The deed transfer is recorded in Deeds Records Philadelphia and Philadelphia County Book MR-7, 717, located in the Philadelphia City Archives; the description of the site is from *Historical,* 2–4; and the price is mentioned in a pamphlet, *Historical Highlights: Frankford Arsenal, Philadelphia, Pennsylvania, U.S. Army Munitions Command, 1816–1973,* found in the library of the U.S. Army Ordnance School, Aberdeen Proving Ground, Aberdeen, Maryland (xeroxed copy provided by Keir Stirling, Ph.D., command historian).

15. *Historical,* 1–4; and Guernsey, "History of Frankford," Papers of the Historical Society, vol. 2, 1906, 3–7.

16. For example, see Col. Ordus C. Horney, "Frankford Arsenal, 1816–1926," *Army Ordnance* 6 (January–February 1926), 233–236.

17. Horney, "Frankford," 233; Collection, 25; and *Historical,* 13–14.

18. Martin, "Samuel Martin"; County Tax Lists 1810 and 1817; and Census 1820.

19. From colonial times until the present an important and substantial black community has existed in Frankford. At times blacks owned property and held skilled occupations. Like blacks in Philadelphia proper, the black community seems to have been seriously affected by nineteenth-century immigration. Although briefly mentioned here, the black community of Frankford deserves more adequate study.

20. For the best study of the Keithian controversy, see J. William Frost, *The Keithian Controversy in Early Pennsylvania* (Norwood, Pa.: Norwood Editions), 1980. For other information on religious congregations and churches, see: Census of 1800, 1810, and 1820 (censuses usually identified Protestant clergymen by congregation); "Historical Report," Hist. Comm.; and List of Members (no date), Frankford Presbyterian church, in the collection of the Office of Presbyterian History.

21. Robert W. Doherty, *The Hicksite Separation: A Sociological Analysis of Religious Schism in Early Nineteenth-Century America* (New Brunswick, N.J.: Rutgers University Press, 1967); Frankford Monthly Meeting, Births and Deaths, Department of Friends' Records, Friends Historical Library, Swarthmore College (microfilm).

22. For the best discussion of Finney's revival and its effects on the Presbyterian church in Philadelphia, see Marion L. Bell, *Crusade in the City: Rivalism in Nineteenth-Century Philadelphia* (Lewisburg, Pa.: Bucknell University Press, 1977); and for an explanation of revivalism and church membership patterns, see Mary P. Ryan, *Cradle of the Middle Class: The Family in Oneida County, New York, 1790–1865* (Cambridge: Cambridge University Press, 1981), especially chapter 3, "The Industrial Era."

23. For a more complete catalog of early nineteenth-century industry in Frankford, see Thompson Westcott and J. Thomas Scharf, *History of Philadelphia* (Philadelphia: L. H. Everts and Co., 1884), 3:2235–2295; "Philadelphia's Tradition of Neighborhoods—Frankford," Hist. Comm.; Martin, "Samuel Martin"; and McConaghy, "Whitaker Mill," 30–34.

24. For an identification of the new immigrant mill owners and work force, see McConaghy, "Whitaker Mill," 48–53; Philip Scranton, *Proprietary Capitalism: The Textile Manufacture at Philadelphia, 1800–1885* (Cambridge: Cambridge University Press, 1983), especially chapter 3; and the censuses of 1830, 1840, 1850, and 1860, all of which are on microfilm in the Philadelphia branch of the National Archives. For a more extended discussion of the polarization of Philadelphia in the nineteenth century, see William W. Cutler III and Howard Gillette, Jr., eds., *The Divided Metropolis: Social and Spatial Dimensions of Philadelphia* (Westport, Conn.: Greenwood Press, 1980).

25. McConaghy, "Whitaker Mill," 38–39; Cynthia J. Shelton, "The Mills of Manayunk: Early Industrialization and Social Conflict in the Philadelphia Region, 1787–1837" (Ph.D. dissertation, University Microfilms International no. DEP-06115, University of California–Los Angeles, 1983).

CHAPTER 2

1. Col. Wadsworth to Capt. Rees, March 10 and May 4, 1819, Frankford Arsenal, Letters Received, Ordnance Office (box 1), 1816–1837, Record Group 156, Entry 1246, National Archives, Philadelphia Branch (hereafter Letters Received, Ord. Off.); Lt. Thomas to Mr. Lee, Esq., Feb. 20, 1821, Frankford Arsenal Letterbook (vol. 1), Aug. 1819–Dec. 1824, Record Group 156, Entry 1234, National Archives, Philadelphia Branch (hereafter Letterbook).

2. "Small arms" is a term used to distinguish pistols, rifles, muskets, and other smaller and lighter firearms from heavier weapons such as cannons, howitzers, mortars, and other artillery.

3. For a more extensive and detailed discussion of the contract system and inspection, see Deyrup, *Arms Makers,* 55–58.

4. A stand of arms was defined as including the weapon itself, a bayonet, and a set of basic tools used to clean and service the weapon.

5. Deyrup, *Arms Makers,* 56.

6. Ibid., 60–67, 91–93; Brooke Hindle and Steven Lubar, *Engines of Change: The American Industrial Revolution, 1790–1860* (Washington, D.C.: Smithsonian Institution Press, 1986), 228–233 and passim.

7. For a discussion of the crucial role of early Philadelphia arms manufacturers, see Smith, *Harpers Ferry Armory,* chapter 2.

8. Emerson Foot to Col. Worth, July 18, 1834, Frankford Arsenal, Letters Received (box 1), 1817–1827, Record Group 156, Entry 1245, National Archives, Philadelphia branch (hereafter Letters Received). For a detailed discussion of the importance of the increased use of gauges in arms production, see Gordon, "Mechanical Reality," 744–778.

9. For one of the sources that convey this impression, see *Historical,* 7–8.

10. B. G. duPont, *Life of Eleuthere Irenee DuPont from Contemporary Sources* (Newark, N.J., 1924), 1:336–337; VanGelder and Schlatter, *History of the Explosive Industry,* 62; and William B. Dixon, "Frankford's Early Industrial Development," *Frankford Historical Society Papers* 2 (1906).

11. Niter and saltpeter are common names for sodium nitrate and potassium nitrate. Along with charcoal and sulfur, they constitute the main ingredients of gunpowder. For the exchange of letters over the duPont holdings for bond surety, see Frankford Arsenal, Letters Received, Ord. Off. (box 1); for the shipments received from the duPont Co., see Lt. Thomas to Col. Bomford, July 11, 1822, Letterbook (vol. 1), and Dec. 27, 1824, and Jan. 15, 1825, Letters Received (box 1).

12. A larger army post would have a variety of officers assigned specific duties, such as paymaster, quartermaster, commissary, and military storekeeper, but at the Frankford Arsenal the commander and sometimes one or two junior officers shared these duties without clear demarcation. For a brief period from about 1838 to 1841 a separate military storekeeper was appointed to the post. This appointment seemed to produce tension since the limits of his duties were never clearly defined. Merrit Roe Smith noted similar circumstances at the Harpers Ferry Arsenal.

13. *Ordnance Manual* (1850), 245–249. A xeroxed copy was provided by John Slonaker, chief librarian at the Institute of Military History, Carlisle Barracks. The title page is missing.

14. Ibid., 248–249.

15. Lt. Thomas to Lt. C. Ward, Oct. 16, 1821, Letterbook (vol. 1); quarterly returns for Third Quarter 1828, Letterbook (vol. 2); and Lt. Thomas to Capt. Beard, Nov. 13, 1820, Letterbook (vol. 2).

16. Lt. Thomas to Mr. Shaw, Nov. 1823, Letterbook (vol. 1). For the best description of the firing mechanism of a musket, see Robert M. Reilly, *United States Military Small Arms, 1816–1865: The Federal Firearms of the Civil War* (Baton Rouge: Eagle Press, 1970), especially the illustrations in the introduction.

17. Col. Walbach to Col. Bomford, June 12, 1832, Letterbook (vol. 2); Lt. Mellon to Col. Bomford, April 23, 1828, Letterbook (vol. 2); and Col. Wadsworth to Capt. Rees, Nov. 1, 1819, Letters Received, Ord. Off. (box 1).

18. Priming tubes were the mechanism used to fire artillery pieces. Their manufacture will be described in more detail in chapter 3.

19. Col. Wadsworth to Capt. Rees, Dec. 17, 1819, Letters Received, Ord. Off. (box 1); Lt. Col. Worth to Col. Bomford, Dec. 1832, Letterbook (vol. 2).

20. George Flegel (master armorer) to Capt. Rees, Jan. 15, 1817, Letters Received (box 1); Col. Bomford to Lt. Col. Worth, July 1834, Letters Received, Ord. Off. (box 2); Lt. Thomas to Col. Bomford, Jan. 20 and Oct. 15, 1823, Letterbook (vol. 1).

21. For evidence of skills among various levels of workers, see: Lt. Thomas to Lt. Hills, April 14, 1820, Letterbook (vol. 1); Lt. Thomas to Lt. (?), April 9, 1823, Letterbook (vol. 1); Lt. Eakins to Col. Bomford, Dec. 5 and Dec. 8, 1827, Letterbook (vol. 2); Lt. Col. Worth to Col. Bomford, Dec. 20, 1832, and Feb. 27, 1833, Letterbook (vol. 2); Capt. Ramsay to Col. Bomford, Dec. 12, 1838, Letterbook (vol. 5); from (?) to Capt. Mordecai, Sept. 3, 1836, Letters Received (box 2); and Col. Bomford to Col. Worth, Aug. 20, 1833, Letters Received, Ord. Off. (box 1). For the description of the explosion in which two laborers were killed, see Lt. Tredwell to Gen. Ripley, Nov. 5, 1861, Letterbook (vol. 13).

22. Lt. Thomas to Col. Bomford, April 11, 1823, Letterbook (vol. 1); Luther Sage, U.S. inspector, to Lt. Thomas, Aug. 5, 1823, and Sept. 14, 1824, Letters Received (box 1); and George Flegel to Lt. T. J. Baird, Jan. 15, 1825, Letters Received (box 1).

23. Col. Walbach to Col. Bomford, Sept. 6, 1831, Letterbook (vol. 2); Lt. Col. Worth to Col. Bomford, March 1833, Letterbook (vol. 2); and Military Storekeeper A. Roumfort to Col. Bomford, Jan. 30, 1839, Letterbook (vol. 5).

24. For evidence of hours worked, see Maj. Laidley to Mr. Thomas Daffin, March 24, 1864, Letterbook (vol. 16). For evidence of wages, see Estimate of Funds for Second Quarter 1833, Letterbook (vol. 2); and Lt. Dearborn to Col. Talcott, July 6, 1846, Letterbook (vol. 7).

25. Ordnance Department Circular Letter, July 1828, Letters Received (box 2).

26. For the preference for enlisted soldiers over hired civilians, see Lt. Eakins to Col. Bomford, Feb. 26, 1828, Letterbook (vol. 2).

27. Lt. Dearborn to A. Webber (M.S.K., Watertown, N.Y.), June 25, 1847, Letters Received (box 5); Capt. Huger to Col. Worth, Sept. 12, 1833, Letters Received (box 2); Col. Worth to Col. Bomford, July 25, 1833, Letterbook (vol. 2); Lt. Mellon to Col. Bomford, Sept. 22, 1828, Letterbook (vol. 2); and Lt. Thomas to Col. Wadsworth, Nov. 13, 1820, Letterbook (vol. 1).

28. For evidence of payment of local taxes, see Lt. Thomas to Mr. Ingersoll, Esq. (U.S. attorney), Feb. 12, 1821, Letterbook (vol. 1); and Capt. Ramsay to Col. Talcott, Dec. 11, 1840, Letterbook (vol. 5).

29. The quotation is from Col. Worth to Col. Bomford, Sept. 6, 1831, Letterbook (vol. 2). For the plans to erect the wall and the neighbors' objections, see Estimated Returns for the Fourth Quarter 1833, Letterbook (vol. 2); Bonney and Bush Co. to M.S.K. Roumfort, March 1838, Letters Received, Ord. Off. (box 1); and copy of letter from Mr. J. Pratt to chief of ordnance, April 1837, Letters Received, Ord. Off. (box 1).

30. The death of Rebecca Deal Murray is announced in Maj. Hagner to Maj. Leslie, Jan. 2, 1860, Letterbook (vol. 12); her admission to the Frankford Presbyterian church is listed in Membership List (1820–26), Frankford Presbyterian Church, Frankford File, Presbyterian Historical Society. The brief snippets of the career of Sgt. John Murray appear in several volumes of the letterbook.

CHAPTER 3

1. Stanley L. Falk, "Soldier-Technologist: Major Alfred Mordecai and the Beginnings of Science in the United States Army" (Ph.D. dissertation, Georgetown University, 1959), 151, 180, 250, and passim.

2. For a recent summary of Jacksonian historiography, see Richard Oestreicher, "Urban Working-Class Behavior and Theories of American Electoral Politics, 1870–1940," *Journal of American History* 74 (March 1988), 1257–1286, especially notes on 1257–1261.

3. Circular, Nov. 9, 1818, Letters Received, Ord. Off. (box 1).

4. Col. Bomford to Lt. Col. Worth, Nov. 10, 1834, Letters Received, Ord. Off. (box 1); Col. Worth to Col. Bomford, Feb. 25, 1833, Letterbook (vol. 2); Second Quarter Estimate, 1828, Letterbook (vol. 2); estimates for year ending June 30, 1849, Letterbook (vol. 7); Office of Commissary General of Subsistence to Lt. Vanness, June 21, 1831, Letters Received (box 1); Circular, Office of Commissary General of Subsistence, July 1, 1833, Letters Received (box 2).

For an extended discussion of the importance of Major Sylvanus Thayer and his reformation of West Point in the development of accounting, accountability, and management in the evolution of industrialization, see Keith Hoskins and Richard Macve, "The Genesis of Accountability: The West Point Connection," *Accounting, Organizations and Society* 13 (1988), 37–73. The thesis presented by Hoskins and Macve is persuasive. They argue that Sylvanus Thayer imposed both discipline and personal and academic accountability on the cadets at West Point. They, in turn, improved the industrial output of arsenals and armories by imposing the same discipline and accountability on workers. For example, in 1833 the barrel welders at the Springfield Armory averaged two hundred barrels per month. By 1842, after the imposition of accountability, the output had increased to four hundred barrels per month. Their extended argument is that industrialization must be seen as more than just mechanization; it is a whole, complex system of organization to be understood. Unfortunately, the case of the Frankford Arsenal does not contribute to their compelling argument. The arsenal did not come to employ a large civilian work force until the Civil War, so there is no opportunity of examining it in transition. Also, while the records of the Frankford Arsenal are fairly extensive, they do not include production records for individual workers. It is not possible, then, to gauge the extent to which workers at the arsenal were managed and held accountable as it is with the workers at Springfield.

5. Lt. Col. Worth to Col. Bomford, Dec. 14, 1832, Letterbook (vol. 2); Lt. Col. Worth to Col. Bomford, Jan. 6, 1833, ibid.; Lt. Col. Worth to Col. Bomford, April 13, 1833, ibid.; Col. Bomford to Capt. Mordecai, Aug. 1, 1836, Letters Received, Ord. Off. (box 1); Col. Bomford to Col. Worth, March 29, 1833, ibid.

6. Falk, "Soldier-Technologist," 234–275; Stirling, *Serving the Line,* 3; *Historical,* 11.

7. Falk, "Soldier-Technologist." The term *soldier-technologist* is unabashedly, but with due deference, borrowed from Stanley Falk for use here.

8. Ibid., 36–86. For another discussion of the importance of West Point to the creation of an industrial system in America, see Peter M. Malloy, "Technical Education and the Young Republic: West Point as America's Ecole Polytechnique, 1802–1833" (Ph.D. dissertation, microfilm no. AAC7615673, Brown University, 1975).

9. Lt. Col. Worth to Col. Bomford, April 13, 1833, and May 12, 1833, Letterbook (vol. 2); Col. Bomford to Col. Worth, May 24, 1833, Letters Received, Ord. Off. (box 1).

10. M.S.K. A. L. Roumfort to duPont Co., June 25, 1840, Letterbook (vol. 5); Capt. Ramsay to Col. Talcott, June 26, 1843, Letterbook (vol. 6); estimates for the year ending 1848, Letterbook (vol. 7); Maj. Ramsay to duPont Co., Jan. 11, 1857, Letterbook (vol. 8); Maj. Hagner to Col. Craig, April 21, 1857, Letterbook (vol. 10); Lt. Treadwell to Col. Ripley, Sept. 4, 1861, Letterbook (vol. 13); Maj. T. S. Laidley to duPont Co., April 16, 1864, Letterbook (vol. 16).

11. Estimates for Fiscal Year 1833, Letterbook (vol. 2); Col. Worth to Col. Bomford, Jan. 16, 1833, Letterbook (vol. 2); estimated returns for Fourth Quarter 1833, Letterbook (vol. 2); Col. Bomford to Col. Worth, Jan. 29, 1833, Letters Received, Ord. Off. (box 1); Col. Bomford to Lt. Col. Worth, May 19, 1854, ibid.

12. Col. Bomford to Col. Worth, March 25, 1833, Letters Received, Ord. Off. (box 1).

13. Col. Bomford to Col. Worth, Jan. 12, 1835, ibid.

14. Falk, "Soldier-Technologist," 7–153 and passim.

15. Ibid., 338–344.

16. Ibid., 344–346; Capt. Mordecai to Col. Bomford, Jan. 8, 1838, Letterbook (vol. 3). Later Mordecai was provided with funds by the Ordnance Department to compile and publish the results of these experiments. They were published as *Reports of Experiments on Gunpowder, Made at Washington Arsenal in 1843 and 1844,* even though some of the experimentation had been done earlier at Frankford.

17. Capt. Mordecai to Col. Bomford, Dec. 4, 1837, Letterbook (vol. 3).

18. Col. Worth to Col. Bomford, Sept. 6, 1831, Letterbook (vol. 2).

19. Col. Bomford to Capt. Mordecai, July 7, 1836, Letters Received, Ord. Off. (box 1); Capt. Mordecai to Col. Bomford, May 7, 1838, Letterbook (vol. 3).

20. Col. Bomford to Capt. Mordecai, Aug. 7, 1838, Letters Received, Ord. Off. (box 1); Falk, "Soldier-Technologist," 257–269, 427, 624, and passim.

21. Col. Bomford to Capt. Ramsay, Dec. 5, 1838, Letters Received, Ord. Off. (box 1); Capt. Ramsay to Col. Bomford, Dec. 1838, Letterbook (vol. 3).

22. Capt. Ramsay to Col. Talcott, May 8, 1844, Letterbook (vol. 6); Capt. Ramsay to Col. Talcott, May 11, 1844, ibid.

23. Sheriff McMichael and Mayor Scott to Capt. Ramsay, May 9, 1844, Letters Received (box 4); McMichael to Capt. Ramsay, May 9, 1844, ibid.

24. The letter, Adjutant General Office to Capt. Ramsay, Aug. 6, 1844, ibid., advised him of the arrival of Company "K"; subsequent letters in Letters Received and the Letterbook over the next year detailed the travails of Company "K." Russell Weigley points out that Capt. Charles F. Smith was an artillery officer. If he arrived with an infantry company, he was in command of a regiment that he did not ordinarily command.

25. A memorandum, Lt. Dearborn to Col. Talcott, June 3, 1846, Letterbook (vol. 7), is a good example of the quantities of material demanded by the war and Lieutenant Dearborn's bemused frustration at the impossibility of some of the demands.

26. Capt. Ramsay to Mr. B. J. Leedman, May 16, 1842, Letterbook (vol. 6); Capt. Ramsay to Col. Talcott, Dec. 26, 1845, Letterbook (vol. 7).

27. Estimates for year ending 1848, Letterbook (vol. 7); estimates for year ending 1849, ibid.

CHAPTER 4

1. Weigley, *History of the United States Army,* chapters 7 and 8.

2. For a discussion of the role of the Ordnance Department in early nineteenth-century technological development, see the introduction by Merrit Roe Smith in Smith, ed., *Military Enterprise,* 1–36.

3. This invention broke a logjam and led to the evolution of small-arms ammunition from percussion caps to center-fire metallic cartridges by 1878. See Reilly, *United States Military,* 18–22.

4. Ibid., 21–22.

5. Oddly, the army resisted the movement toward rapid-firing arms on the grounds that such firing made battles uncontrollable and expensive.

6. For an extensive discussion of conservatism in the Ordnance Department, see Michael S. Raber,

"Conservative Innovators, Military Small Arms, and Industrial History at Springfield Armory, 1794–1918," *Journal of the Society for Industrial Archaeology* 14 (1988), 1–21.

7. Keir B. Stirling, *Serving the Line with Excellence: The Development of the U.S. Army Ordnance Corps* (Washington, D.C.: U.S. Government Printing Office, 1987), 19–20.

8. Agreement between Mr. Joseph Deal and U.S. Gov't., March 16, 1850, Letters Received (box 5).

9. The quotation is from a letter, Major Ramsay to (?), April 4, 1851, Letterbook (vol. 8).

For some of the important works on the social impact of industrialization in the Delaware Valley area in the first half of the nineteenth century, see: Shelton, "Mills"; William Sullivan, *The Industrial Workers in Pennsylvania, 1800–1840* (Harrisburg: Pennsylvania Historical and Museum Commission, 1955); Anthony F. C. Wallace, *Rockdale: The Growth of an American Village in the Early Industrial Revolution* (New York: Alfred A. Knopf, 1978); and Laurie, *Working People.*

10. The first quotation is from Estimates of Funds, First Quarter of 1851, Letterbook (vol. 8), and the second quotation is from Agreement between Mr. Joseph Deal and U.S. Gov't., May 26, 1851, Letters Received (box 5).

11. Stirling, *Serving the Line,* 20.

12. Ibid., 21.

13. Sept. 17, 1851, Letterbook (vol. 8).

14. *Historical,* 21. In spite of his importance to the development of munition and armament technology, Peter V. Hagner remains an enigmatic figure. Other than his birth in Washington, D.C., as the son of a vestryman in the Episcopal church, his graduation from West Point, and his career as an ordnance officer, little is available in the way of biographical information.

15. Maj. Hagner to Col. Craig, Sept. 20, 1851, Letterbook (vol. 8); Col. Craig to Maj. Hagner, Feb. 2, 1852, Letters Received, Ord. Off. (box 6); *Historical,* 21–22; Maj. Hagner to Col. Craig, April 20, 1852, Letterbook (vol. 8).

16. Maj. Hagner to Col. Craig, Dec. 12, 1851, Letterbook (vol. 8); Maj. Hagner to Col. Craig, Dec. 31, 1851, ibid.; Col. Craig to Maj. Hagner, Feb. 2, 1852, Letters Received, Ord. Off. (box 6); Maj. Hagner to Col. Craig, March 29, 1852, Letterbook (vol. 8).

17. Maj. Hagner to Col. Craig, July 23, 1852, Letterbook (vol. 8). The sources do not record which machine came from which specific arsenal, nor where the machines were built. They only mention that Robert Bo[u]lton's cap machine had been patented while he was at the Watervliet Arsenal.

18. Within the last thirty years, historians employing new methodologies and exploring new sources have begun to write a new social history of workers, their workplace, and their communities. Led by European historians such as E. J. Hobsbawm, George Rude', and E. P. Thompson, we have learned to understand inarticulate people who left no written records, but who have spoken to us through their actions and other mediums.

But just as we have been informed, so also we have been misled. The major thesis of Thompson's work is that by the 1830s the English workers had come to identify themselves as a class opposed to others, and that they had developed "class consciousness." Beginning with Herbert Gutman, "Work, Culture, and Society in Industrializing America, 1815–1891," in *Work, Culture, and Society in Industrializing America* (New York: Random House, 1977), other historians of the United States have followed in Thompson's footsteps. In his excellent work on the industrialization of the shoe industry in Lynn, Massachusetts, Alan Dawley describes the evolution of shoemaking from an artisanal skill to an industrial occupation with the subsequent "deskilling" and degradation of once independent craftsmen. He portrays them in Thompsonesque fashion as having developed a class consciousness focused around the organization of the Knights of St. Crispin, only to see this nascent class consciousness dissipate in the heroic fervor of the Civil War. In a similar study of textile workers in Philadelphia, Bruce Laurie argues for the appearance of class consciousness in the form of the General Trades Union and the general strike of 1837. Like other American historians, Laurie is hard-pressed to account for the rapid disappearance of class consciousness after 1837. To do so he borrows from trendy sociological studies to posit the splintering of class solidarity into five personality types. Others such as Anthony F. C. Wallace and Paul Johnson have identified evangelization as the death knell of class consciousness in places as diverse as the Brandywine Valley of Chester County, Pennsylvania, and the Erie Canal city of Rochester, New York. More recently, in her work on the textile industry of Manayunk, Cynthia J. Shelton searches for class consciousness only to find workers more interested in substantive economic goals than in class issues. These scholars begin with the assumption that the coming of mechanized industry resulted in the dimunition of skills without ever examining or defining the meaning of skills, as applied to either craftsmen or industrial workers. Within the last several years Robert B. Gordon and Michael S. Raber, have challenged the assumption that mechanization automatically deskilled workers. Having searched so hard for class consciousness, some historians have failed to examine the actual relationship between work and workers.

19. The career of William Adams is detailed in various volumes of the Letterbook of the Frankford Arsenal, especially vol. 8.

20. The career of William Pigott is also detailed in Letterbook (vol. 8); the quotation is from Maj. Hagner to Col. Craig, Dec. 28, 1853.

21. George Wright is also described in Letterbook (vol. 8) and in *Historical,* 19–21, 26–27, and 95–97. The quotation is from Maj. Hagner to Col. Craig, Dec. 28, 1853, Letterbook (vol. 8).

22. Like the others, George Esher is also described in Letterbook (vol. 8).

23. The cases of George Esher and George Wright support the argument advanced by Raber, "Conservative Innovators," and Gordon, "Material Evidence": that increased production of small arms resulted not solely from increased mechanization, but from the increased skill of workers using increasingly numerous and sophisticated gauges to hand-finish machined work.

24. Maj. Hagner to Col. Craig, May 10, 1852, Letterbook (vol. 8); Maj. Hagner to Mr. Perkins, May 24, 1852, ibid.; Maj. Hagner to Col. Craig, Feb. 1, 1853, ibid.

25. Messrs. Dougherty and Thomas to Capt. Reno, Dec. 30, 1854, Letters Received (box 6); Col. Craig to Maj. Hagner, June 17, 1857, Letters Received, Ord. Off. (box 4).

26. Robert Bolton to Maj. Hagner, Oct. 16, 1857, Letters Received (box 7).

27. Lt. Dearborn to 2d Auditor, Treas. Dept., Letterbook (vol. 7); the letter is undated.

28. Maj. Ramsay to Col. Craig, March 3, 1851, Letterbook (vol. 8); Maj. Hagner to Col. Craig, Jan. 2, 1852, ibid.; Maj. Hagner to Col. Craig, Sept. 25, 1856, Letterbook (vol. 10).

29. *The Ordnance Manual for the Use of Officers of the United States Army,* 3d ed., (Philadelphia: J. B. Lippincott and Co., 1864), 300–303. These pages, like those of the 1850 *Manual* cited earlier, were xeroxed copies provided by John Slonaker.

30. Ibid., 300–301.

31. Ibid., 302.

32. Maj. Hagner to Col. Craig, Nov. 7, 1853, Letterbook (vol. 9).

33. Col. Craig to Maj. Hagner, Nov. 9, 1853, Letters Received, Ord. Off. (box 3).

34. Col. Craig to Maj. Hagner, Nov. 1, 1853, Letters Received, Ord. Off. (box 3); Maj. Hagner to Col. Craig, Dec. 2, 1854, Letterbook (vol. 9). Raber, in "Conservative Innovators," argues that the increased use of gauging as much as mechanization was responsible for the attainment of uniformity and interchangeability of parts by the mid-1850s in small-arms production.

35. Hagner, a clever man as well as an innovator, took the precaution of having the machine photographed when it was completed. Unfortunately neither the photograph nor any drawing of the machine has been found.

36. Maj. Hagner to Col. Craig, March 3, 1856, Letterbook (vol. 9); Maj. Hagner to Maj. Ramsay, Dec. 29, 1857, Letterbook (vol. 10).

37. Maj. Hagner to Col. Craig, March 3, 1856, Letterbook (vol. 10); Maj. Hagner to Col. Craig, Jan. 15, 1857, ibid.

38. Maj. Hagner to Col. Craig, Aug. 11, 1855, Letterbook (vol. 9); Maj. Hagner to E. Remington

and Sons, Jan. 30, 1856, ibid.; Maj. Hagner to Col. Craig, Feb. 27, 1856, ibid.; Maj. Hagner to E. Remington and Sons, April 29, 1856, and May 16, 1856, ibid.; Col. Craig to Maj. Hagner, April 22, 1857, Letters Received, Ord. Off. (box 4); Maj. Hagner to Col. Craig, March 25, 1858, Letterbook (vol. 10).

39. Smith, "Army Ordnance," 39. Hoke, "Ingenious Yankees," argues that in the 1850s the federal armories "in the closed environment" stagnated, while private manufacturers took the lead in industrial innovations. He may be correct in the case of the federal armories at Harpers Ferry and Springfield, but not in respect to the Frankford Arsenal.

40. For evidence of the change in the focus of the Arsenal, see Letterbook (vol. 10).

41. Maj. Hagner to Col. Craig, June 30, 1860, Letterbook (vol. 12); Stirling, *Serving the Line,* 22.

42. For a discussion of the concept of "mechanical reality," see Gordon, "Mechanical Reality," 744–77.

CHAPTER 5

1. Stirling, *Serving the Line,* 25–26; *Historical,* 27–28.

2. Stirling, *Serving the Line,* 26.

3. Keir Stirling suggests that Gorgas, a Northerner by birth, may have been motivated as much by personal ambition as by conviction when he accepted command of the Confederate ordnance department. Stirling lists several instances when Gorgas had been denied recognition and promotion by Col. Henry Knox Craig and the ordnance office.

4. The quotations are from Capt. Maynardier to Simon Cameron, April 17, 1861, Letterbook (vol. 12); Lt. Treadwell to Col. Ripley, May 30, 1861, ibid.; Lt. Treadwell to Col. Ripley (telegram), Sept. 5, 1861, ibid.; and Lt. Treadwell to Gen. Ripley, Sept. 14, 1861, Letterbook (vol. 13). Why the charges against Perkins and Bolton appeared in a New York newspaper is not explained. No records exist of the hearing before the United States commissioner in the papers of the U.S. Circuit Court, Eastern District for Pennsylvania.

5. Paul Johnson, *Shopkeeper's Millennium,* 38–61.

6. Gen. Ripley to Lt. Treadwell, Nov. 8, 1861, Letters Received, Ord. Off. (box 5); Lt. Treadwell to Gen. Ripley, Dec. 21, 1861, Letterbook (vol. 14); Lt. Treadwell to Sellers Company, Jan. 6, 1862, ibid.; Maj. Laidley to Gen. Ripley, Feb. 18, 1862, ibid.

Eventually the Ordnance Department rewarded Lieutenant Treadwell for his herculean labors. In the spring of 1862 the chief of ordnance promoted Treadwell to the rank of captain, transferred him to Hilton Head, South Carolina (just occupied by Union forces), and gave him the title "commander of ordnance in the South—a title perhaps meant to rankle Josiah Gorgas.

7. Lt. Treadwell to Col. Maynardier, Nov. 5, 1861, Letterbook (vol. 13); Gen. Ripley to Capt. Rodman and Maj. Laidley, Letters Received, Ord. Off. (box 6); copy of letter, "Wm. Woodington, Paul St., Frankford," addressed to "The President of the United States," March 11, 1863, Letters Received (box 12). This letter complained of discounted "Certificates of Indebtedness."

8. Maj. Laidley to Mr. Thomas Daffin, April 24, 1864, Letterbook (vol. 16); Maj. Laidley to Sarah Dobson, June 25, 1864, ibid.; Jabez Gill, "Toolmaker and Machinist," "Bridesburg," to Maj. Laidley, May 17, 1864, Letters Received (box 14); Maj. Laidley to Jabez Gill, June 16, 1864, Letterbook (vol. 16).

9. Maj. Laidley to Gen. Thomas, Adjutant General, July 23, 1864, Letterbook (vol. 16). Within a year an explosion killed Jacob Waltzhauer. His family does not appear in the vicinity of the arsenal in the Census of 1870.

10. Letter to Maj. Laidley, June 12, 1864, Letters Received (box 13).

11. Capt. Maynardier to Col. Craig, April 16, 1861, Letterbook (vol. 12); Lt. Treadwell to Col. Ripley, May 30, 1861, ibid.; Lt. Treadwell to Col. Ripley, June 6, 1861, Letterbook (vol. 13).

12. Callum, *Biographical Register,* 2:116–117.

13. Maj. Laidley to E. C. Seamen and R. Dolby Company, March 1862, Letterbook (vol. 14); Gen. Ripley to Maj. Laidley, April 25, 1862, and July 3, 1862, Letters Received, Ord. Off. (box 5); Maj. Laidley to Maj. Whiteley (New York Arsenal), May 23, 1862, Letterbook (vol. 14).

14. *Historical,* 28–29; Maj. Laidley to Stidham Company (Philadelphia), July 24, 1862, Letterbook (vol. 15); Maj. Laidley to Messrs. Aug. Viele Company (West Troy, N.Y.), Aug. 8, 1862, ibid.; Maj. Laidley to Morris and Tasker Company (Philadelphia), Aug. 19, 1862, ibid. Being a dedicated scientist, Laidley even blew up a building to test the new construction.

15. Stirling, *Serving the Line,* 28–29. Ramsay's tenure lasted less than one year. He rapidly became disenamored with the machinations of Stanton and Captain Balch, and in September 1864 he was relieved of command.

16. *Historical,* 32–35 (including a detailed drawing of the column design); Maj. Laidley to Phoenix Iron Company, Nov. 11, 1862, Letterbook (vol. 15).

17. Maj. Laidley to Corliss Company, Aug. 14, 1863, Letterbook (vol. 15); Putnam Machine Company to Capt. Benet, Dec. 3, 1864, Letters Received (box 15). Some work has been done on the relationship between architecture and other activities such as teaching (see Cutler and Gillette, *Divided Metropolis*), but not on the relationship between the new nineteenth-century industrial workplace and the organization of work within it, although Daniel T. Rogers, *The Work Ethic in Industrial America* (Chicago: University of Chicago Press, 1974), does approach the subject in his discussion of Frederick W. Taylor; see chapter 2.

18. *Historical,* 28–33 Maj. Laidley to Mr. James Green (New York), Nov. 11, 1863, Letterbook (vol. 15).

19. *Historical,* 36–37.

20. Gen. Ramsay to Maj. Laidley, Oct. 3, 1863, Letters Received, Ord. Off. (box 6). Hoke, "Ingenious Yankees," argues that private industry rather than the Ordnance Department advanced the creation of an industrial system of manufacture during the 1850s. This letter undermines Hoke's argument.

21. Laidley's letters to the Phoenix Iron Company, dated March 21, 1864, April 13, 1864, and May 25, 1864, are found in Letterbook (vol. 16). The Phoenix Iron Company's letters to Laidley, dated June 6, 1864, June 7, 1864, July 2, 1864, and Aug. 11, 1864, are from Letters Received (boxes 13 and 14). The special order relieving Laidley of command is Gen. Ramsay to Maj. Laidley, Aug. 11, 1864, Letters Received, Ord. Off. (box 6).

22. Stirling, *Serving the Line,* 32–33.

23. Frankford Arsenal, Monthly Return of Hired Men, Oct. 1864–Dec. 1867, Record Group 156, Entry 1255, 346–355 and 523–533, in the Philadelphia branch of the National Archives (hereafter Monthly Returns).

24. Ibid.

25. Ibid.

26. Monthly Returns; Censuses of 1860 and 1870.

27. Monthly Returns; Frankford Arsenal, Manufacturing Reports, 1867–1895, Record Group 156, Entry 1263, also in the Philadelphia branch of the National Archives.

CHAPTER 6

1. "Returns from U.S. Military Posts, 1800–1916," microfilm no. 676, Frankford Arsenal, Pa., roll 373, June 1833–December 1876, National Archives Microfilm Publications (privately owned); Monthly Returns.

2. "Population Schedule of the Ninth Census of the United States, 1870," microfilm roll 1410; First Enumeration, Philadelphia County, Philadelphia City, Wards 23 and 25 (hereafter Census of 1870). This microfilmed census schedule is also housed in the Philadelphia branch of the National Archives.

3. The Census of 1870 also identifies an A. G. Pigott, a thirty-five-year-old clerk, but it also lists his wife as Isabella, his two-year-old daughter as Estelle, and Hugh Pigott, a twenty-two-year-old machinist, as members of the same household. This could be a garbled listing of two different brothers and their households, or a double enumeration.

For the best work on wages and standard of living in late nineteenth-century Philadelphia, see Eudice Glassberg, "Work, Wages and the Cost of Living: Ethnic Differences and the Poverty Line, Philadelphia, 1880," *Pennsylvania History* 66 (January 1979), 17–58.

4. Census of 1870.

5. Census of 1870; Jabez Gill's home and property are described in *City Atlas of Philadelphia,* vol. 4, 25th ward (Philadelphia: G. M. Hopkins Co., 1875), plates F and G.

6. Census of 1870.

7. Ibid.

8. Ibid.

9. Ibid. For the best discussion of Victorian attitudes about female employment, see Nancy F. Cott, *The Bonds of Womenhood: "Woman's Sphere" in New England, 1780–1835* (New Haven: Yale University Press, 1977), especially the introduction and 21–56.

10. Census of 1870.

11. Ibid.

12. For the best estimates on the amount of income required to sustain a household in 1870, see Glassberg, "Work, Wages and the Cost of Living," 17–58. Glassberg estimated that a family of four (a mother, father, and two children) needed $643.60 in annual income to live at a subsistence level; this is almost identical to estimates by the Massachusetts Bureau of Labor in 1880, and to Census of 1870 and Monthly Returns.

13. Census of 1870 and Monthly Returns.

14. Ibid.

15. Ibid.

16. In 1868 both the commander of the Frankford Arsenal, Lt. Col. Stephen Vincent Benet, and the first officer, Capt. Joseph P. Farley, were Roman Catholics and attended the local Catholic church, All Saints. On January 3, 1869, Captain Farley's daughter Eleanor was baptized in this church. The baptism is listed in "All Saints Register of Baptisms." It, along with the registers of deaths and marriages, was made available by the current pastor, the Rev. Joseph P. Kennedy.

17. Census of 1870 and Monthly Returns.

18. This conclusion substantiates other recent studies that interpret industrial or commercial employment of adolescents as part of familial or paternal strategy. See Hareven, *Family Time and Industrial Time,* especially chapters 3, 5, and 8; and Alan Stanley Horlick, *Country Boys and Merchant Princes: The Social Control of Young Men in New York* (Lewisburg, Pa.: Bucknell University Press, 1975). Horlick argues that successful fathers actually preferred initiating their sons into business over continuing their formal education.

19. By 1870 a horsecar line connected Bridesburg with the northern boundary of Philadelphia's urban area, but the cost of the fare (five cents) would have prohibited commuting for most laborers.

20. Harry C. Silcox, "Henry Disston's Model Industrial Community: Nineteenth-Century Paternalism in Tacony, Philadelphia," *Pennsylvania Magazine of History and Biography* 114 (October 1990), 483–515. See also Silcox, *A Place to Live and Work: The Henry Disston Saw Works and the Tacony Community of Philadelphia* (University Park: Pennsylvania State University Press, 1994).

21. Hopkins, *City Atlas,* plate G.

22. Census of 1870.

23. Census of 1870, Manufacturing Schedule, Book 11, 82nd District, 25th Ward (xeroxed copy among the collection of the Philadelphia Social History Project held by Van Pelt Library, University of Pennsylvania); Ronald R. Gurczynski, *Bridesburg: Yesterday . . . Today* (Bridesburg Bicentennial Committee, 1976), 6.

24. Census of 1870; *Centennial History, First Presbyterian Church at Bridesburg,* a pamphlet in the collection of the Office of Presbyterian History. The church was organized in 1837, and services were held in Alfred Jenks's home. In 1868 the congregation erected the present building, and Alfred Jenks contributed thirty-two thousand dollars toward the construction. The quotation is from W. B. Heydrick and appeared in Philadelphia *Sunday Bulletin,* March 11, 1951, p. 4. Mr. Heydrick's family was among the oldest in Bridesburg. The land on which Alfred Jenks built his factory had originally been owned by the Heydrick family.

25. The Census of 1870 identifies the birthplace of Lennig's workers. Unconfirmed local legend holds that Lennig actively recruited these workers and "brought them over." Before the end of the century large numbers of Polish immigrants began to arrive, joining the other ethnic groups already living there. Among the works on bi-occupationalism and partible inheritance in late nineteenth-century southwestern Germany are Wolfgang Zorn, "Bayerns Gewerbe, Handel und Verkehr (1806–1907)" and "Die Sozialentwicklungder," in *Handbuch der Bayerischen Geschite,* ed. Max Spindler (Munich: Beck, 1967), 4:782–841, 846–877; and Mack Walker, *German Home Towns, Community, State, and General Estate, 1648–1871* (Ithaca: Cornell University Press, 1971), especially chapter 6.

26. Gurczynski, *Bridesburg,* 11–12.

27. Census of 1870; Monthly Returns; Hopkins, *City Atlas.*

28. Membership lists exist only for the First Presbyterian church at Bridesburg. Between 1837 and 1871, 325 people were admitted to membership. About half of those admitted either moved, died, or were dismissed in any five-year period. At no time did the congregation exceed approximately one hundred and twenty people. All Saints Parish, like most Roman Catholic churches in the nineteenth century, kept no official membership list; but figures in the parish registers of births, marriages, and deaths suggest a Catholic congregation about equal to the Presbyterian one. If these rough calculations are extended to other churches, certainly less than half of the inhabitants of Bridesburg were church members.

29. Census of 1870.

30. Regina Radocaj et al., *All Saints Church, Bridesburg, Pa., 1860–1985,* 15–19. This is a memorial book published on the 125th anniversary of the parish. A copy was given to me by the pastor, the Rev. Joseph P. Kennedy.

Bibliography

UNPUBLISHED SOURCES

I. Public Records

A. Microfilm

U.S. Department of Commerce. Bureau of the Census. "United States Census of Population: Third Census, 1810." Pennsylvania, Philadelphia County, M 252, Roll 56, Philadelphia Branch of the National Archives.

U.S. Department of Commerce. Bureau of the Census. "United States Census of Population: Fourth Census, 1820." Pennsylvania, Vol. 14, M 33, Roll 109, Philadelphia Branch of the National Archives.

U.S. Department of Commerce. Bureau of the Census. "United States Census of Population: Fifth Census, 1830." Pennsylvania, Vol. 16, Roll 158, Philadelphia Branch of the National Archives.

U.S. Department of Commerce. Bureau of the Census. "United States Census of Population: Sixth Census, 1840." Pennsylvania, Vol. 26, Roll 491, Philadelphia Branch of the National Archives.

U.S. Department of Commerce. Bureau of the Census. "United States Census of Population: Seventh Census, 1850." Pennsylvania, Vol. 45, Roll 824, Philadelphia Branch of the National Archives.

U.S. Department of Commerce. Bureau of the Census. "United States Census of Population: Eighth Census, 1860." Pennsylvania, Philadelphia, 23rd Ward, Vol. 63, Roll 1174, Philadelphia Branch of the National Archives.

U.S. Department of Commerce. Bureau of the Census. "United States Census of Population: Ninth Census, 1870." First Enumeration, Pennsylvania, Philadelphia, 23rd and 25th Wards, Vol. 62, Roll 1410. Philadelphia Branch of the National Archives.

U.S. Department of Commerce. Bureau of the Census. "United States Census of Population: Ninth Census, 1870." Second Enumeration, Pennsylvania, Philadelphia, 23rd and 25th Wards, Vol. 62, Roll 1438. Philadelphia Branch of the National Archives.

U.S. Department of Treasury. "Direct Tax of 1798." Pennsylvania, Philadelphia, M 372, Roll 5, Philadelphia Branch of the National Archives.

U.S. General Service Administration. National Archives and Record. Records of the Chief of Ordnance. Record Group 156. "Returns from U. S. Military Posts, 1800–1916," Philadelphia Branch of the National Archives.

B. Non-Microfilm

Archives of the City of Philadelphia. Department of Records. Record Group 207. "Frankford Borough Ordnances, 1800–1846." 2 vols.

Archives of the City of Philadelphia. Department of Records. Record Group 1.9. "Philadelphia County Tax, Oxford Township, 1800, 1803–1813, 1815–1817." 3 vols.

U.S. Department of Commerce. Bureau of the Census. "United States Census of Population: Ninth Census, 1870." Manufacturing Schedule, 25th Ward, 82d District, Book 11. A photocopy in the collection of the Philadelphia Social History Project, Van Pelt Library, University of Pennsylvania.

U.S. General Service Administration. National Archives and Record Service. Records of the Chief of Ordnance. Record Group 156. Entry 1234. Frankford Arsenal. "Letterbook." 20 vols. Philadelphia Branch of the National Archives.

U.S. General Service Administration. National Archives and Record Service. Records of the Chief of Ordnance. Record Group 156. Entry 1245. Frankford Arsenal. "Letters Received." 198 Boxes. Philadelphia Branch of the National Archives.

U.S. General Service Administration. National Archives and Record Service. Records of the Chief of Ordnance. Record Group 156. Entry 1246. Frankford Arsenal. "Letters Received, Ordnance Office." 11 Boxes. Philadelphia Branch of the National Archives.

U.S. General Service Administration. National Archives and Record Service. Records of the Chief of Ordnance. Record Group 156. Entry 1263. Frankford Arsenal. "Manufacturing Reports, 1867–1895." 1 vol. Philadelphia Branch of the National Archives.

U.S. General Service Administration. National Archives and Record Service. Records of the Chief of Ordnance. Record Group 156. Entry 1255. Frankford Arsenal. "Monthly Returns of Hired Men, 1864–1867." 1 vol. Philadelphia Branch of the National Archives.

U.S. General Service Administration. National Archives and Record Service. Records of the Chief of Ordnance. Record Group 336. Entry 21. "Ordnance Department File, 1797–1894." Box 147. National Archives, Washington, D.C.

II. Private Records

All Saints Parish. "Register of Baptisms, 1860–1893." 1 vol. All Saints Parish Office, Philadelphia.

All Saints Parish. "Register of Deaths, 1860–1888." 1 vol. All Saints Parish Office, Philadelphia.

All Saints Parish. "Register of Marriages, 1860–1894." 1 vol. All Saints Parish Office, Philadelphia.

First Presbyterian Church of Bridesburg. "Record of Church Members, 1837–1871." 1 vol. First Presbyterian Church of Bridesburg, Philadelphia.

Frankford Monthly Meeting. "Record of Births, 1797–1830." Department of Friends' Records, Friends Historical Library, Swarthmore College, Swarthmore, Pennsylvania.

Frankford Monthly Meeting. "Record of Membership, 1818–1831." Department of Friends' Records, Friends Historical Library, Swarthmore College, Swarthmore, Pennsylvania.

Frankford Presbyterian Church. "Record of Church Members, 1808–1830." 1 vol. Frankford File, Office of Presbyterian History, Philadelphia.

Philadelphia County. "Philadelphia County Tax List, Oxford Township, 1800." 1 vol. Pennsylvania Historical Society. Philadelphia.

Saint Joachim Parish. "Register of Baptisms, 1848–1872." 1 vol. Saint Joachim Parish Office, Philadelphia.

Saint Joachim Parish. "Register of Deaths, 1848–1872." 1 vol. Saint Joachim Parish Office, Philadelphia.

Saint Joachim Parish. "Register of Marriages, 1848–1872." 1 vol. Saint Joachim Parish Office, Philadelphia.

III. Manuscript Collections

Duffield, Charles, H. "The Old Swedes Mill: Its Water Rights and Surrounding Industries." June 1907. Historical Society of Pennsylvania, Philadelphia (typescript).

Historical Society of Pennsylvania. "Description of Land in Oxford Township." Norris MSS. undated. Philadelphia.

Historical Society of Pennsylvania. "Early Land Surveys of Pennsylvania and Delaware, 1671–1681." Miscellaneous Papers of Philadelphia County, 1671–1738. Philadelphia.

Historical Society of Pennsylvania. "Frankford Creek Farm." 1806 survey by John Hills, Logan MSS. Philadelphia.

Historical Society of Pennsylvania. "Frankford Creek, land on, survey of." undated. Miscellaneous Papers of Philadelphia County, 1671–1738. Philadelphia.

Historical Society of Pennsylvania. "Frankford Road." March 10, 1729/30, surveyed by Isaac Norris, Logan Papers. Philadelphia.

Historical Society of Pennsylvania. "Petition of the Inhabitants of Frankford for a Footpath." 1705. Proud MSS. Philadelphia.

Historical Society of Pennsylvania. "Petition to Governor William Markham." Miscellaneous Papers of Philadelphia County, 1671–1738. Philadelphia.

Historical Society of Pennsylvania. "Report to the Provincial Council on the Frankford Road, Sept. 29, 1701." Logan Papers. Philadelphia.

IV. Dissertations and Theses

Cesari, G. S. "American Arms Making Machine Tool Development." Ph.D. dissertation, University of Pennsylvania, 1970.

Ezell, Edward C. "The Development of Artillery for the United States Land Service before 1861." M.A. thesis, University of Delaware, 1963.

Falk, Stanley. "Soldier-Technologist: Alfred Mordecai and the Beginning of Science in the United States Army." Ph.D. dissertation, Georgetown University, 1959.

Fones-Wolf, Kenneth. "Trade Union Gospel: Protestantism and Labor in Philadelphia, 1865–1915." Ph.D. dissertation, Temple University, 1985.

Hoke, Donald. "Ingenious Yankees: The Rise of the American System of Manufacture in the Private Sector." Ph.D. dissertation, University of Wisconsin–Madison, 1984.

Shelton, Cynthia J. "The Mills of Manayunk: Early Industrialization and Social Conflict in the Philadelphia Region, 1787–1837." Ph.D. dissertation, University of California–Los Angeles, 1983.

V. Maps

Philadelphia Historical Commission. "Map of Philadelphia, 1774." Map File. Philadelphia.
Philadelphia Historical Commission. "Map of Philadelphia, 1809." Map File. Philadelphia.
Philadelphia Historical Commission. "Map of Philadelphia, 1843." Map File. Philadelphia.
Philadelphia Historical Commission. "Map of Philadelphia County, March 19, 1816." Map File. Philadelphia.
Philadelphia Historical Commission. "Thomas Holme's Map of Philadelphia, 1687." Map File. Philadelphia.

PUBLISHED SOURCES

I. Secondary Works

Adams, Donald R., Jr. *Wage Rates in Philadelphia, 1790–1830.* Dissertations in American History. New York: Arno Press, 1975.
Bell, Marion L. *Crusade in the City: Revivalism in Nineteenth-Century Philadelphia.* Lewisburg, Pa.: Bucknell University Press, 1977.
Blodget, Lorin. *The Industries of Philadelphia as Shown by the Manufacturing Census of 1870, Compared with 1860 and Estimates for 1875 and 1876.* Philadelphia: Collins, 1877.
Chandler, Alfred Dupont, Jr. *The Visible Hand: The Managerial Revolution in American Business.* Cambridge, Mass.: Belknap Press, 1977.
Cresson, William Penn. *James Monroe.* Chapel Hill: University of North Carolina Press, 1946.
Cullum, George W. *Biographical Register of the Officers and Graduates of the U.S. Military Academy.* 8 vols. 3d edition. New York: Houghton Mifflin, 1940.
Cutler, William W., III, and Gillette, Howard, Jr., eds. *The Divided Metropolis: Social and Spatial Dimensions of Philadelphia, 1800–1975.* Westport, Conn.: Greenwood Press, 1980.
Daly, John. *Descriptive Inventory of the Archives of the City and County of Philadelphia.* Philadelphia: City of Philadelphia Department of Records, 1970.
Dawley, Alan. *Class and Community: The Industrial Revolution in Lynn.* Cambridge, Mass.: Harvard University Press, 1976.
Deyrup, Felicia Johnson. *Arms Makers of the Connecticut Valley: A Regional Study of the Economic Development of the Small Arms Industry, 1798–1870.* Smith College Studies in History, vol. 33. Northampton, Mass., 1948.
Doherty, Robert W. *The Hicksite Separation: A Sociological Analysis of Religious Schism in Early Nineteenth-Century America.* New Brunswick, N.J.: Rutgers University Press, 1967.
Dublin, Thomas. *Women at Work: The Transformation of Work and Community in Lowell, Massachusetts, 1826–1860.* New York: Columbia University Press, 1979.
duPont, B. G. *The Life of Eleuthere Irenee duPont from Contemporary Correspondence.* 5 vols. Newark, Del., 1924.
Frost, J. William, compiler. *The Keithian Controversy in Early Pennsylvania.* Norwood, Pa.: Norwood Editions, 1980.
Gurczynski, Ronald R. *Bridesburg: Yesterday . . . Today.* Published by the Bridesburg Bicentennial Committee, 1976.

Hackley, Col. Frank W. *A History of Modern U.S. Military Small Arms Ammunition.* New York: Macmillan, 1967.

Hareven, Tamara K. *Family Time and Industrial Time: The Relationship Between the Family and Work in a New England Industrial Community.* New York: Cambridge University Press, 1981.

Hershberg, Theodore, ed. *Philadelphia: Work, Space, Family, and Group Experience in the Nineteenth Century.* New York: Oxford University Press, 1981.

Hindle, Brooke, and Lubar, Steven. *Engines of Change: The American Industrial Revolution, 1790–1860.* Washington, D.C.: Smithsonian Institution Press, 1986.

Historical and Archeological Survey of Frankford Arsenal, Philadelphia, Pennsylvania. 2 vols. Prepared for the Department of the Army, Baltimore District Corps of Engineers, by John Milner Associates. Philadelphia, 1979.

Hobsbawn, Eric J. *Labouring Men: Studies in the History of Labour.* London: Weidenfeld & Nicolson, 1968.

Hopkins, G. M. *City Atlas of Philadelphia, By Ward, Complete in 7 Volumes.* Philadelphia, 1876.

Horlick, Alan Stanley. *Country Boys and Merchant Princes: The Social Control of Young Men in New York.* Lewisburg, Pa.: Bucknell University Press, 1975.

Horowitz, Daniel. *Morality of Spending: Attitudes Toward the Consumer Society in America, 1875–1940.* Baltimore: Johns Hopkins University Press, 1985.

Hounshell, David A. *From the American System to Mass Production, 1800–1932.* Baltimore: Johns Hopkins University Press, 1984.

Huntington, Samuel P. *The Soldier and the State: The Theory and Politics of Civil-Military Relations.* Cambridge, Mass.: Belknap Press, 1957.

Johnson, Amandus. *The Swedes on the Delaware, 1638–1664.* Philadelphia: Swedish Colonial Society, 1915.

Johnson, Paul E. *A Shopkeeper's Millennium: Society and Revivals in Rochester, New York, 1815–1837.* New York: Hill & Wang, 1978.

Laurie, Bruce. *The Working People of Philadelphia, 1800–1850.* Philadelphia: Temple University Press, 1980.

Leiby, Adrian Coulter. *The Early Dutch and Swedish Settlers of New Jersey.* Princeton, N.J.: Van Nostrand, 1967.

Lemon, James T. *The Best Poor Man's Country: A Geographical Study of Early Southeastern Pennsylvania.* New York: Norton and Co., 1976.

Macfarlane, Alan. *The Origins of English Individualism: The Family, Property and Social Transition.* New York: Cambridge University Press, 1978.

McGaw, Judith. *Most Wonderful Machine: Mechanization and Social Change in Berkshire Paper-Making, 1801–1885.* Princeton: Princeton University Press, 1987.

Montgomery, David. *Worker's Control in America.* New York: Cambridge University Press, 1979.

More, Charles. *Skill and the English Working Class.* London: Macmillan, 1980.

The Ordnance Manual for the Use of the Officers of the United States Army. 3d edition. Philadelphia: J. B. Lippincott and Co., 1864.

Radocaj, Regina, et al. *All Saints Church, Bridesburg, Pa., 1860–1985.* Bridesburg, Pa.: By the parish anniversary committee, 1985.

Randall, Willard S., and Solomon, Stephen D. *Building 6: The Tragedy of Bridesburg.* Boston: Little, Brown & Co., 1975.

Regulations for the Inspection of Small Arms. Washington, D.C., 1823.

Reilly, Robert M. *United States Military Small Arms, 1816–1865: The Federal Firearms of the Civil War.* Baton Rouge: Eagle Press, 1970.

Rude', George. *The Crowd in History: A Study of Popular Disturbances in France and England, 1730–1848.* Revised edition. London: Lawrence & Wishart, 1981.

Ryan, Mary P. *Cradle of the Middle Class: The Family in Oneida County, New York, 1790–1865.* Cambridge: Cambridge University Press, 1981.

Scranton, Philip. *Proprietary Capitalism: The Textile Manufacture at Philadelphia, 1800–1885.* Cambridge: Cambridge University Press, 1983.

Smith, Merrit Roe. *Harpers Ferry Armory and the New Technology: The Challenge of Change.* Ithaca: Cornell University Press, 1977.

Smith, Merrit Roe, ed. *Military Enterprise and Technological Change: Perspectives on the American Enterprise.* Cambridge: MIT Press, 1985.

Spindler, Max. *Handbuch der Bayerischen Geschichte.* Munich: Beck, 1967.

Stirling, Keir B. *Serving the Line with Excellence: The Development of the U.S. Army Ordnance Corps.* Washington, D.C.: U.S. Government Printing Office, 1987.

Sullivan, William. *The Industrial Worker in Pennsylvania, 1800–1840.* Harrisburg: Pennsylvania Historical and Museum Commission, 1955.

Thompson, E. P. *The Making of the English Working Class.* Vintage edition. New York: Alfred A. Knopf and Random House, 1966.

VanCreveld, Martin. *Technology and War: From 2000 B.C. to the Present.* New York: The Free Press, 1989.

Vandiver, Frank E. *Ploughshares into Swords: Josiah Gorgas and Confederate Ordnance.* Austin: University of Texas Press, 1952.

VanGelder, Arthur Pine, and Schlatter, Hugo. *History of the Explosive Industry.* New York: Macmillan, 1927.

Wainwright, Nicholas B. *Philadelphia in the Romantic Age of Lithography.* Philadelphia: Historical Society of Pennsylvania, 1958.

Wallace, Anthony F. C. *Rockdale: The Growth of an American Village in the Early Industrial Revolution.* New York: Alfred A. Knopf, 1978.

Wallace, Anthony F. C. *The Social Context of Innovation.* Princeton: Princeton University Press, 1982.

Weigley, Russell F. *History of the United States Army.* Enlarged edition. Bloomington: Indiana University Press, 1967.

Weslager, Clinton A. *Dutch Explorers, Traders and Settlers in the Delaware Valley, 1609–1664.* Philadelphia: University of Pennsylvania Press, 1961.

Westcott, Thompson, and Scharf, J. Thomas. *History of Philadelphia.* 3 vols. Philadelphia: L. H. Everts and Co., 1884.

Wolf, Stephanie G. *Urban Village, Population, Community, and Family Structure in Germantown, Pennsylvania, 1683–1800.* Princeton: Princeton University Press, 1976.

Wrightson, Keith, and Levine, David. *Poverty and Piety in an English Village: Terling, 1525–1700.* New York: Academic Press, 1979.

II. Periodicals

Cooper, Carolyn C. "A Whole Battalion of Stockers: Thomas Blanchard's Production Line and Hand Labor at Springfield Armory." *Journal of the Society for Industrial Archaeology* 4, no. 1 (1988), 37–56.

Dixon, W. B. "Frankford's Early Industrial Development." *Frankford Historical Society Papers* 2 (1906), 1–4, 62–64.

Ericksen, Eugene P., and Yancey, William L. "Work and Residence in Industrial Philadelphia." *Journal of Urban History* 5 (February 1979), 147–182.

Glassberg, Eudice. "Work, Wages and the Cost of Living: Ethnic Differences and the Poverty Line, Philadelphia, 1880." *Pennsylvania History* 66 (January 1979), 17–58.

Gordon, Robert B. "Material Evidence of Manufacturing Methods Used in Springfield Armory." *Journal of the Society for Industrial Archaeology* 14, no. 1 (1988), 57–75.

Gordon, Robert B. "Who Turned the Mechanical Ideal into Mechanical Reality?" *Technology and Culture* 29 (October 1988), 744–778.

Hornsey, Odus C., Col. "1816–Frankford Arsenal–1926." *Army Ordnance* 6 (January–February 1926), 233–236.

McConaghy, Mary. "The Whitaker Mill, 1813–1843: A Case Study of Workers, Technology, and Community in Early Industrial Philadelphia." *Pennsylvania History* 51 (January 1984), 30–53.

Montgomery, David. "The Working Classes of the Pre-Industrial American City, 1780–1830." *Labor History* 9 (Winter 1968), 3–22.

Oestreicher, Richard. "Urban Working-Class Political Behavior and Theories of American Electoral Politics, 1870–1940." *Journal of American History* 74 (March 1988), 1257–1286.

Pi-Sunyer, O., and DeGregori, Thomas. "Cultural Resistance to Technological Change." *Technology and Culture* 5 (1964), 247–253.

Pi-Sunyer, O., and DeGregori, Thomas. "Technology, Traditionalism, and Military Establishments." *Technology and Culture* 7 (1966), 402–407.

Raber, Michael S. "Conservative Innovators, Military Small Arms, and Industrial History at Springfield Armory, 1794–1918." *Journal of the Society for Industrial Archaeology* 14, no. 1 (1988), 1–21.

Rosenberg, Nathan. "Technological Change in the Machine Tool Industry, 1840–1910." *Journal of Economic History* 23 (1963), 414–443.

Silcox, Harry C. "Henry Disston's Model Industrial Community: Nineteenth-Century Paternalism in Tacony, Philadelphia." *Pennsylvania Magazine of History and Biography* 114 (October 1990), 483–515.

Simler, Lucy. "The Landless Worker: An Index of Economic and Social Change in Chester County, Pennsylvania, 1750–1820." *Pennsylvania Magazine of History and Biography* 114 (April 1990), 163–199.

Smedley, Caroline. "Historical Sketches of Frankford Meeting." *Frankford Historical Society Papers* 2, no. 5 (1916), 219–225.

Thompson, E. P. "Time, Work-Discipline, and Industrial Capitalism." *Past and Present* 38 (1967), 56–97.

Ulle, Robert. "Frankford, Philadelphia: The Development of a Nineteenth-Century Urban Black Community." *Pennsylvania Heritage* (December 1977), 2–8.

Uselding, Paul J. "Technical Progress at the Springfield Armory, 1820–1850." *Explorations in Economic History* 9 (1972), 291–316.

MISCELLANEOUS SOURCES

"Historical Highlights: Frankford Arsenal, Philadelphia, Pennsylvania, 19137, U.S. Army Munitions Command, 1816–1973; Established on the 27th of May 1816." A xeroxed copy of a memorial pamphlet provided by Keir B. Stirling, command historian, Ordnance Corps, Aberdeen Proving Ground, Aberdeen, Md.

Office of Presbyterian History. "Centennial History, First Presbyterian Church at Bridesburg."
      Pamphlet in the Bridesburg file.
Office of Presbyterian History. "First Presbyterian Church of Bridesburg, 150th Anniversary."
      Pamphlet in the Bridesburg file.
Philadelphia Sunday Bulletin, March 11, 1951. A clipping privately owned by Frank
      Devlin, Marlton, N.J.

Index